Variations on Split Plot and Split
Block Experiment Designs

THE WILEY BICENTENNIAL–KNOWLEDGE FOR GENERATIONS

Each generation has its unique needs and aspirations. When Charles Wiley first opened his small printing shop in lower Manhattan in 1807, it was a generation of boundless potential searching for an identity. And we were there, helping to define a new American literary tradition. Over half a century later, in the midst of the Second Industrial Revolution, it was a generation focused on building the future. Once again, we were there, supplying the critical scientific, technical, and engineering knowledge that helped frame the world. Throughout the 20th Century, and into the new millennium, nations began to reach out beyond their own borders and a new international community was born. Wiley was there, expanding its operations around the world to enable a global exchange of ideas, opinions, and know-how.

For 200 years, Wiley has been an integral part of each generation's journey, enabling the flow of information and understanding necessary to meet their needs and fulfill their aspirations. Today, bold new technologies are changing the way we live and learn. Wiley will be there, providing you the must-have knowledge you need to imagine new worlds, new possibilities, and new opportunities.

Generations come and go, but you can always count on Wiley to provide you the knowledge you need, when and where you need it!

WILLIAM J. PESCE
PRESIDENT AND CHIEF EXECUTIVE OFFICER

PETER BOOTH WILEY
CHAIRMAN OF THE BOARD

Variations on Split Plot and Split Block Experiment Designs

WALTER T. FEDERER
Cornell University
Departments of Biological Statistics and Computational Biology and Statistical Sciences
Ithaca, NY

FREEDOM KING
Cornell University
Department of Biological Statistics and Computational Biology
Ithaca, NY

WILEY-INTERSCIENCE
A JOHN WILEY & SONS, INC., PUBLICATION

Copyright 2007 by John Wiley & Sons, Inc. All rights reserved

Published by John Wiley & Sons, Inc., Hoboken, New Jersey
Published simultaneously in Canada

No part of this publication may be reproduced, stored in a retrieval system, or transmitted in any form or by any means, electronic, mechanical, photocopying, recording, scanning, or otherwise, except as permitted under Section 107 or 108 of the 1976 United States Copyright Act, without either the prior written permission of the Publisher, or authorization through payment of the appropriate per-copy fee to the Copyright Clearance Center, Inc., 222 Rosewood Drive, Danvers, MA 01923, (978) 750-8400, fax (978) 750-4470, or on the web at www.copyright.com. Requests to the Publisher for permission should be addressed to the Permissions Department, John Wiley & Sons, Inc., 111 River Street, Hoboken, NJ 07030, (201) 748-6011, fax (201) 748-6008, or online at http://www.wiley.com/go/permission.

Limit of Liability/Disclaimer of Warranty: While the publisher and author have used their best efforts in preparing this book, they make no representations or warranties with respect to the accuracy or completeness of the contents of this book and specifically disclaim any implied warranties of merchantability or fitness for a particular purpose. No warranty may be created or extended by sales representatives or written sales materials. The advice and strategies contained herein may not be suitable for you situation. You should consult with a professional where appropriate. Neither the publisher nor author shall be liable for any loss of profit or any other commercial damages, including but not limited to special, incidental, consequential, or other damages.

For general information on our other products and services or for technical support, please contact our Customer Care Department within the United States at (800) 762-2974, outside the United States at (317) 572-3993 or fax (317) 572-4002.

Wiley also publishes its books in a variety of electronic formats. Some content that appears in print may not be available in electronic format. For more information about Wiley products, visit our web site at www.wiley.com.

Library of Congress Cataloging-in-Publication Data

Federer, Walter Theodore, 1915-Variations on split plot and split block experiment
 designs / Walter T. Federer, Freedom King.
   p. cm.
 Includes bibliographical references and index.
 ISBN-13: 978-0-470-08149-5 (acid-free paper)
 ISBN-10: 0-470-08149-X (acid-free paper)
1. Experimental design. 2. Blocks (Group theory) I. King, Freedom, 1955- II. Title.

QA279.F427 2007
519.5'7–dc22   2006049672

10 9 8 7 6 5 4 3 2 1

To Edna, my lovely wife and helpmate

To Fidela, my loved wife and our sons
Emmanuel, Willy, Fabrice and Yves.

Contents

Preface   xiii

Chapter 1. The standard split plot experiment design   1
     1.1. Introduction   1
     1.2. Statistical design   3
     1.3. Examples of split-plot-designed experiments   6
     1.4. Analysis of variance   9
     1.5. F-tests   12
     1.6. Standard errors for means and differences between means   14
     1.7. Numerical examples   16
     1.8. Multiple comparisons of means   23
     1.9. One replicate of a split plot experiment design and missing observations   26
     1.10. Nature of experimental variation   28
     1.11. Repeated measures experiments   29
     1.12. Precision of contrasts   29
     1.13. Problems   31
     1.14. References   32
     Appendix 1.1. Example 1.1 code   34
     Appendix 1.2. Example 1.2 code   36

Chapter 2. Standard split block experiment design   39
     2.1. Introduction   39
     2.2. Examples   41
     2.3. Analysis of variance   43
     2.4. F-tests   44
     2.5. Standard errors for contrasts of effects   45
     2.6. Numerical examples   46
     2.7. Multiple comparisons   52
     2.8. One replicate of a split block design   53
     2.9. Precision   53

2.10. Comments	54
2.11. Problems	55
2.12. References	55
Appendix 2.1. Example 2.1 code	56
Appendix 2.2. Example 2.2 code	56
Appendix 2.3. Problems 2.1 and 2.2 data	60

Chapter 3. Variations of the split plot experiment design — 61

3.1. Introduction	61
3.2. Split split plot experiment design	62
3.3. Split split split plot experiment design	67
3.4. Whole plots not in a factorial arrangement	73
3.5. Split plot treatments in an incomplete block experiment design within each whole plot	74
3.6. Split plot treatments in a row-column arrangement within each whole plot treatment and in different whole plot treatments	75
3.7. Whole plots in a systematic arrangement	76
3.8. Split plots in a systematic arrangement	77
3.9. Characters or responses as split plot treatments	77
3.10. Observational or experimental error?	79
3.11. Time as a discrete factor rather than as a continuous factor	80
3.12. Inappropriate model?	86
3.13. Complete confounding of some effects and split plot experiment designs	90
3.14. Comments	91
3.15. Problems	91
3.16. References	93
Appendix 3.1. Table 3.1 code and data	94

Chapter 4. Variations of the split block experiment design — 97

4.1. Introduction	97
4.2. One set of treatments in a randomized complete block and the other in a Latin square experiment design	98
4.3. Both sets of treatments in split block arrangements	100
4.4. Split block split block or strip strip block experiment design	100
4.5. One set of treatments in an incomplete block design and the second set in a randomized complete block design	106
4.6. An experiment design split blocked across the entire experiment	107
4.7. Confounding in a factorial treatment design and in a split block experiment design	108
4.8. Split block experiment design with a control	111
4.9. Comments	114

4.10. Problems	114
4.11. References	115
Appendix 4.1. Example 4.1 code	115

Chapter 5. Combinations of SPEDs and SBEDs — 120

5.1. Introduction	120
5.2. Factors A and B in a split block experiment design and factor C in a split plot arrangement to factors A and B	120
5.3. Factor A treatments are the whole plot treatments and factors B and C treatments are in a split block arrangement within each whole plot	125
5.4. Factors A and B in a standard split plot experiment design and factor C in a split block arrangement over both factors A and B	127
5.5. A complexly designed experiment	130
5.6. Some rules to follow for finding an analysis for complexly designed experiments	135
5.7. Comments	138
5.8. Problems	139
5.9. References	139
Appendix 5.1. Example 5.1 code	139
Appendix 5.2. Example 5.2 data set, code, and output	144

Chapter 6. World records for the largest analysis of variance table (259 lines) and for the most error terms (62) in one analysis of variance — 147

6.1. Introduction	147
6.2. Description of the experiment	148
6.3. Preliminary analyses for the experiment	152
6.4. A combined analysis of variance partitioning of the degrees of freedom	157
6.5. Some comments	163
6.6. Problems	163
6.7. References	163
Appendix 6.1. Figure 6.1 to Figure 6.6	164

Chapter 7. Augmented split plot experiment design — 169

7.1. Introduction	169
7.2. Augmented genotypes as the whole plots	170
7.3. Augmented genotypes as the split plots	174
7.4. Augmented split split plot experiment design	176
7.5. Discussion	180
7.6. Problems	180
7.7. References	181

Appendix 7.1. SAS code for ASPED, genotypes as whole plots,
Example 7.1   182
Appendix 7.2. SAS code for ASPEDT, genotypes as split plots,
Example 7.2   185
Appendix 7.3. SAS code for ASSPED, Example 7.3   186

Chapter 8. Augmented split block experiment design   **188**
    8.1. Introduction   188
    8.2. Augmented split block experiment designs   188
    8.3. Augmented split blocks for intercropping experiments   193
    8.4. Numerical example 8.1   194
    8.5. Comments   197
    8.6. Problems   197
    8.7. References   198
    Appendix 8.1. Codes for numerical Example 8.1   198

Chapter 9. Missing observations in split plot and split block experiment designs   **202**
    9.1. Introduction   202
    9.2. Missing observations in a split plot experiment design   203
    9.3. Missing observations in a split block experiment design 204
    9.4. Comments   204
    9.5 Problems   204
    9.6. References   206
    Appendix 9.1. SAS code for numerical example in Section 9.2   206
    Appendix 9.2. SAS code for numerical example in Section 9.3   209

Chapter 10. Combining split plot or split block designed experiments over sites   **213**
    10.1. Introduction   213
    10.2. Combining split plot designed experiments over sites   213
    10.3. Combining split block designed experiments over sites   217
    10.4. Discussion   219
    10.5. Problems   219
    10.6. References   219
    Appendix 10.1. Example 10.1   219
    Appendix 10.2. Example 10.2   229

Chapter 11. Covariance analyses for split plot and split block experiment designs   **239**
    11.1. Introduction   239
    11.2. Covariance analysis for a standard split plot design   240
    11.3. Covariance analysis for a split block experiment design 250
    11.4. Covariance analysis for a split split plot experiment design 255

11.5. Covariance analysis for variations of designs	259
11.6. Discussion	260
11.7. Problems	260
11.8. References	262
Appendix 11.1. SAS code for Example 11.1	263
Appendix 11.2. SAS code for Example 11.2	264
Appendix 11.3. SAS code for Example 11.3	265

Index   **267**

Preface

When it comes to designed experiments, researchers often end up creating complex designs without having sufficient analytical expertise to handle. Researchers in plant breeding, animal science, health sciences and so forth, come to statistical consulting with data from rather very complex designs from time to time. Unfortunately, statistical courses taken by these researchers may not have covered these sophisticated designs. To make matters even more severe, there is an alarming shortage of textbooks covering complex designs. To help alleviate the analytical challenges of researchers dealing with complex designs, we have decided to write this book and we do hope that it will be helpful to a lot of researchers. Understanding and mastery of the designs covered here, assume a prior exposure to the basic experimental designs such as: one-way completely randomized design, completely randomized factorial experiment designs, randomized complete blocks with one or more factors, incomplete blocks, row-column designs, Latin-square designs and so forth. These basic designs are easy to analyze since one is dealing with one experimental error given one has a single level of randomization of the treatment combinations between the levels of various factors to the experimental units. Nonetheless, this type of randomization might be rather simplistic and inappropriate depending on the existing experimental conditions along with the constraints imposed by limited resources. As a result, the experimenter might be forced to have different randomizations and therefore experimental units of unequal sizes at different levels of randomization, to overcome logistical and/or technological constraints of an experiment. This opens up a class of more complex designs called split plot designs or split block designs with at least two types of experimental errors. In either case, several variations can occur with a possibility of a further partitioning of the experimental units, leading to smaller and smaller experimental units paralleled with more error terms used to test the significance of various factors' effects. Furthermore, an experiment design might consist of a combination of these two types of designs, along with treatments arranged following the basic designs for some of the factors under investigation. A textbook on variations of split plot and split block designs points in the right direction by addressing the urgent need of researchers dealing with complex designs for which no reference is available to the best of our knowledge. We have encountered a few researchers in this type of situation through our statistical consulting activities. We are therefore convinced that this book will be a valuable resource not only to researchers but also to instructors teaching experiment designs courses. It is also important to adequately equip graduate students with the important skills in complex designs for a better readiness to real life situation challenges as far as designed experiments are concerned. Another important innovation of this textbook consists of tackling the issue of error reduction through blocking, analysis of covariance, or both. While blocking relatively homogeneous experimental units into groups might help reduce substantially the experimental error, there are situations where it is neither sufficient by itself nor feasible at all. Thus, use of available auxiliary information on the experimental units has proven to significantly reduce the experimental error through analysis of covariance. Analysis of covariance enables one to better control the experimental error when covariates are judiciously chosen. We have added a chapter on analysis of covariance to specifically provide researchers with helpful analytical tools needed when dealing with covariates in complex designs. Accompanying data sets can be found on the Wiley FTP site. ftp://ftp.wiley.com/public/sci_tech_med/split_plot/

WALTER T. FEDERER
FREEDOM KING
May 2006

CHAPTER 1

The Standard Split Plot Experiment Design

1.1. INTRODUCTION

Prior to starting the topic of this book, it was deemed advisable to present some design concepts, definitions, and principles. Comparative experiments involve a number, v, treatments (factors) where a *treatment* is an item of interest to the experimenter. A treatment could be a medical treatment, a drug application, a level of a factor (amount of a drug, fertilizer, insecticide, etc.), a genotype, an agricultural practice, a marketing method, a teaching method, or any other item of interest. The selection of the v treatments for an experiment is known as the *treatment design*. The selection of an appropriate treatment design is a major element for the success of an experiment. It may include checks (standards, placebos) or other *points of reference*. The treatments may be all combinations of two or more factors and this is known as a *factorial arrangement* or *factorial treatment design*. A subset of a factorial is denoted as a *fractional replicate* of a factorial.

The arrangement of the treatments in an experiment is known as the *experiment design* or the *design of the experiment*. The term experimental design is of frequent use in statistical literature but is not used here. There are many types of experiment designs including: unblocked designs, blocked designs (complete blocks and incomplete blocks), row-column experiment designs, row-column designs within complete blocks, and others. Tables of designs are available in several statistical publications. However, many more experiment designs are available from a software package such as GENDEX (2005). This package obtains a randomized form of an experiment design and the design in variance optimal or near optimal.

There are three types of units to be considered when conducting an experiment. These are the observational unit, the sample or sampling unit, and the experimental

---

Variations on Split Plot and Split Block Experiment Designs, by Walter T. Federer and Freedom King
Copyright © 2007 John Wiley & Sons, Inc.

unit (Federer, 1991, Chapter 7). The *observational unit* is the smallest unit for which a response or measurement is obtained. A population or distribution is composed of *sample units or sampling units*. The *experimental unit* is the smallest amount of experimental material to which one treatment is applied. In many experiments, these three types of units are one and the same. In other cases, they may all be different. For example, suppose a treatment is a teaching method taught to a group of thirty students. The experimental unit is the group of thirty students for the period of time used to evaluate a teaching method. The sampling unit is the student, from a population of all students, for which inferences are to be made about this teaching method. Suppose that several examinations are given during the period of time the method is applied, the result from each examination is an observation or response and the observational unit is one examination from one student. In some investigations like sampling for water quality, obtaining a measurement on produce for a genotype from a plot of land measuring 1 m by 10 m (an experimental unit), etc., the sampling units are undefined.

Fisher (1966) presented three principles of experiment design. These are *local control (blocking, stratification), replication,* and *randomization*. Owing to random fluctuations of responses in any experiment or investigation, there is variation. The variation controlled should not be associated or interacting with treatment responses. For example, if an animal dies during the course of conducting an experiment and the death is not caused by the treatment, it should be considered as a missing observation and not as a zero response. Blocking (stratification) or local control is used to exclude extraneous variation in an experiment not associated with treatment effects. The blocking should be such as to have maximum variation among blocks and minimum variation within blocks. This makes for efficient experimentation and reduces the number of replicates (replications) needed for a specified degree of precision for treatment effects.

To reduce the effect of the variation in an experiment on measuring a treatment effect, the sample size or the number of replicates needs to be increased. Replication allows for an estimate of the random variation. Replication refers to the number of experimental units allocated to a particular treatment. The variation among the experimental units, eliminating treatment and blocking effects, is a measure of experimental variation or error. The number of replications should not be confused with the number of observations. For example, in a nutrition study of several regimes with an experimental unit consisting of one animal, weekly measurements (observations) may be taken on the weight of the animal over a 6-month period. These week-by-week measurements do not constitute replications. The number of replications is determined by the number of experimental units allocated to one treatment and not by the number of observations obtained.

Randomization is necessary in order to have a valid estimate of an error variance for comparing differences among treatments in an experiment. Fisher (1966) has defined a *valid estimate of an error variance or mean square* as one which contains all sources of variation affecting treatment effects except those due to the treatments themselves. This means that the estimated variance should be among experimental units treated alike and not necessarily among observations.

An appropriate response model needs to be determined for each experiment. It is essential to determine the pattern of variation in an experiment or investigation and not assume that one response model fits all experiments for a given design. With the availability of computers, *exploratory model selection* may be utilized to determine variation patterns in an experiment (Federer, 2003). The nature of the experiment design selected and the variation imposed during the conduct of an experiment determine the variation pattern. The conduct of an experiment or investigation is a part of the design of the experiment or investigation. This fact may be overlooked when selecting a response model equation for an experiment. For example, a randomized complete block design may be selected as the design of the experiment. Then, during the course of conducting the experiment, a part of the replicate of the experiment is flooded with water. This needs to be considered as a part of the design of the experiment and may be handled by setting up another block, using a covariate, or missing experimental units. This would not be the response model envisioned when the experiment design was selected. Or, it may be that the experimenter observed an unanticipated gradient in some or all of the blocks. A response model taking the gradients within blocks into consideration should be used in place of the model presumed to hold when the experiment was started. More detail on exploratory model selection may be found in Federer (2003).

For further discussion of the above, the reader is referred to Fisher (1966) and Federer (1984). The latter reference discusses a number of other principles and axioms to consider when conducting experiments.

An analysis of variance is considered to be a partitioning of the total variation into the variation for each of the sources of variation listed in a response model. An F-test is not considered to be a part of the analysis of variance as originally developed by Sir Ronald A. Fisher. Statistical publications often consider an F-test as part of the analysis of variance. We do not, as variance component estimation, multiple range tests, or other analyses may be used in connection with an analysis of variance. Some experimenters do consider the term analysis of variance to be a misnomer. A better term may be a partitioning of the total variation into its component parts or simply variation or variance partitioning.

1.2. STATISTICAL DESIGN

The standard split plot experiment design (SPED) discussed in several statistics textbooks has a two-factor factorial arrangement as the treatment design. One factor, say A with a levels, is designed as a randomized complete block design with r complete blocks or replicates. The experimental unit, the smallest unit to which one treatment is applied, for the levels of factor A treatments is called a *whole plot experimental unit* (wpeu). Then each wpeu is divided into b *split plot experimental units* (speus) for the b levels of the second factor, say B. Note that either or both factors A and B could be in a factorial arrangement or other treatment design rather than a single factor. A schematic layout of the standard SPED is shown below.

Standard split plot design with r replicates, a levels of factor A, and b levels of factor B

Replicate	1	2	3	...	r
Whole plot factor A	1 2 ... a	1 2 ... a	1 2 ... a	...	1 2 ... a
Split plot factor B	1 1 ... 1	1 1 ... 1	1 1 ... 1		1 1 ... 1
	2 2 ... 2	2 2 ... 2	2 2 ... 2		2 2 ... 2
	......	......	......		......
	$b\,b...b$	$b\,b...b$	$b\,b...b$		$b\,b...b$

The a levels of factor A are randomly and independently allocated to the a wpeus within *each* of the r complete blocks or replicates. Then within *each* wpeu, the b levels of factor B are independently randomized. There are r independent randomizations for the a levels of factor A and ra independently assigned randomizations for b levels of factor B. The fact that the *number of randomizations and the experimental units are* different for the two factors implies that each factor will have a separate error term for comparing effects of factor A and effects of factor B.

Even though the standard SPED has the whole plot factor A treatments in a randomized complete block design, any experiment design may be used for the factor A. For example, a completely randomized experiment design, a Latin square experiment design, an incomplete block experiment design, or any other experiment design may be used for the whole plot treatments. These variations are illustrated in Chapter 3.

The three steps in randomizing a plan for a standard or basic split plot experiment design consisting of $r = 5$ blocks (replicates), $a = 4$ levels of whole plot factor A, and $b = 8$ levels of split plot factor B are shown below:

Step 1: Divison of the experimental area or material into five blocks

STATISTICAL DESIGN

Step 2: Randomizaton of four levels of whole plot factor A to each of five blocks

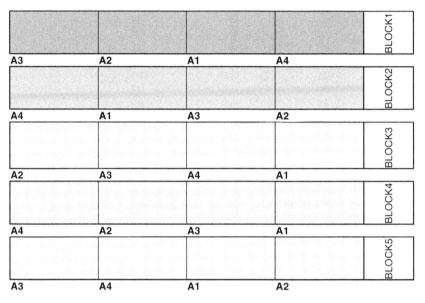

Step 3: Randomization of eight levels of split plot factor B within each level of whole plot factor A

B1	B2	B7	B2	
B4	B3	B8	B4	
B5	B4	B4	B7	
B3	B5	B2	B5	
B6	B1	B5	B8	
B8	B6	B3	B1	
B7	B8	B6	B6	
B2	B7	B1	B3	BLOCK1
A3	A2	A1	A4	
B7	B6	B2	B5	
B2	B1	B3	B4	
B4	B4	B5	B2	
B6	B3	B7	B8	
B3	B7	B8	B3	
B8	B2	B1	B6	
B1	B5	B6	B7	
B5	B8	B4	B1	BLOCK2
A4	A1	A3	A2	
B4	B7	B1	B6	
B6	B8	B2	B1	
B1	B2	B4	B3	
B8	B6	B3	B5	
B5	B3	B7	B4	
B7	B1	B8	B8	
B3	B5	B6	B7	
B2	B4	B5	B2	BLOCK3
A2	A3	A4	A1	

B3	B7	B5	B8	
B8	B6	B2	B5	
B5	B2	B3	B4	
B6	B8	B4	B1	
B1	B4	B7	B3	
B7	B5	B6	B6	
B4	B1	B8	B7	
B2	B3	B1	B2	BLOCK4
A4	A2	A3	A1	
B3	B1	B7	B1	
B1	B6	B2	B7	
B5	B2	B3	B4	
B6	B7	B4	B8	
B8	B4	B5	B3	
B7	B5	B6	B6	
B4	B8	B1	B5	
B2	B3	B8	B2	BLOCK5
A3	A4	A1	A2	

If an experiment design involving blocking is used for the b split plot treatments, factor B, should be *within each* whole-plot-treatment wpeu, as this facilitates the statistical analysis for an experiment as orthogonality of effects is maintained. If the experiment design for the split plot factor B treatments is over levels of the whole plot treatments within one complete block, confounding of effects is introduced and the statistical analysis becomes more complex (Federer, 1975). This may not be a computational problem as available statistical software packages can be written to handle this situation. However, the confounding of effects reduces the precision of contrasts and estimates of effects.

1.3. EXAMPLES OF SPLIT-PLOT-DESIGNED EXPERIMENTS

Example 1—A seed germination test was conducted in a greenhouse on $a = 49$ genotypes of guayule, the whole plots (factor A), with four seed treatments (factor B) applied to each genotype as split plot treatments (Federer, 1946). The wpeu was a greenhouse flat for one genotype and 100 seeds of each of the four seed treatments (factor B) were planted in a flat, as more information on seed treatment than on genotype was desired and this fitted into the layout more easily than any other arrangement. The speu consisted of 1/4 of a greenhouse flat in which 100 seeds were planted. The 49 genotypes were arranged in a triple lattice incomplete block experiment design with $r = 6$ complete blocks and with an incomplete block size of $k = 7$ wpeus. The four seed treatments were randomly allocated to the four speus in a flat, that is, within each genotype wpeu. The data for eight of the 49 genotypes in three of the six replicates are given as Example X-1 of Federer (1955) and as Example 1.2. The whole plot treatments, 49 genotypes, are considered to be a random sample of genotypes from a population of genotypes, that is, they are considered to be random effects whereas the seed treatments are fixed effects as these are the only ones of interest.

Example 2—Example X-2 of Federer (1955) contains the yield data for $b = 6$ genotypes which are corn double crosses. The data are from two of the twelve districts set up for testing corn hybrids in Iowa. The $a = 2$ districts are the whole plots, and the six corn double crosses, the split plot treatments, are arranged in a randomized complete block design within each district. The yield data (pounds of ear corn) arranged systematically are given below:

District 1, A					
Double-cross, factor B	Replicate 1	Replicate 2	Replicate 3	Replicate 4	Total
1-1	34.6	33.4	36.5	33.0	137.5
2-2	34.5	39.1	35.4	35.6	144.6
4-3	30.1	30.8	35.0	33.3	129.2
15-45	31.3	29.3	29.7	33.2	123.5
8-38	32.8	35.7	36.0	34.0	138.5
7-39	30.7	35.5	35.3	30.6	132.1
Total	194.0	203.8	207.9	199.7	805.4

District 2, A					
Double-cross, factor B	Replicate 1	Replicate 2	Replicate 3	Replicate 4	Total
1-1	33.1	24.6	33.8	34.6	126.1
2-2	46.4	36.9	36.3	45.3	164.9
4-3	32.3	38.7	37.5	37.6	146.1
15-43	37.5	39.2	39.1	34.1	149.9
8-38	31.2	40.8	46.1	44.1	162.2
7-39	35.8	38.2	38.8	39.6	152.4
Total	216.3	218.4	231.6	235.3	901.6

Example 3—Cochran and Cox (1957), page 300, present the data for an SPED with $a = 3$ recipes, the whole plots (factor A), for chocolate cakes baked at $b = 6$ temperatures, the split plots (factor B). The response was the breaking angle of the cake. Enough batter for one recipe was prepared for the six cakes to be baked at the six temperatures. That is, the wpeu was one batter for six cakes. The three recipes were arranged in a randomized complete block design with $r = 15$ replicates.

Example 4—Federer (1955), page 26 of the Problem Section, presents the data for an SPED with $a = 2$ whole plot treatments (factor A) of alfalfa or no alfalfa and $b = 5$ split plot treatments of bromegrass strains. The bromegrass strains were intercropped (mixed together) with the alfalfa and no alfalfa (See Federer,

1993, 1999). The whole plot treatments were arranged in a randomized complete block design with $r = 4$ replicates. The dry weights (grams) of hay arranged systematically are:

Bromegrass strain, factor B	Replicate 1 Factor A		Replicate 2 Factor A		Replicate 3 Factor A		Replicate 4 Factor A	
	alfalfa	alone	alfalfa	alone	alfalfa	alone	alfalfa	alone
a	730	786	1004	838	871	1033	844	867
b	601	1038	978	1111	1059	1380	1053	1229
c	840	1047	1099	1393	938	1208	1170	1433
d	844	993	990	970	965	1.308	1111	1311
e	768	883	1029	1130	909	1247	1124	1289

Example 5—Das and Giri (1979), page 150, present an example of three varieties forming the whole plots and $b = 4$ manurial treatments forming the split plots in an SPED with $r = 4$ replications.

Example 6—Gomez and Gomez (1984), page 102, give a numerical example of six levels of nitrogen applications forming the whole plots and $b = 4$ rice varieties forming the split plots in an SPED with $r = 3$ replications.

Example 7—Raghavarao (1983), page 255, presents a numerical example where the whole plots were $a = 3$ nitrogen levels and the $b = 4$ split plot treatments were insecticides in an SPED with $r = 4$ replications.

Example 8—Leonard and Clark (1938), Chapter 21, give a numerical example of a split plot experiment design with $a = 10$ maize hybrids as the whole plots of 36 hills (3 plants per hill). The wpeus were divided into thirds with 12 hills making up the speu. The $b = 3$ split plot treatments were seeds from the three generations F1, F2, and F3. Two replicates were used and the response was the yield of ear corn.

Example 9—In a setting other than agriculture, three types of schools (public, religious, and private) were the whole plots. Four types of teaching methods formed the split plots. This arrangement was replicated over r school districts. The response was the average score on standardized tests.

Example 10—Two types of shelters (barn and outdoor) were the whole plot factor A treatments and two types of shoes for horses were used as the factor B split plot treatments. There were to be $r = 5$ sets (replicates) of four horses used. Two horses, wpeu, of each set would be kept in a barn and two would be kept outdoors. One horse, speu, had one type of shoe and the second horse received the other type of shoe. The response was length of time required before reshoeing a horse was required.

Example 11—In a micro-array experiment, the two whole plot treatments were methods one and two. The two split plot treatments were red color-label 1 and green color-label 2 for method 1 and were green color-label 1 and red color-label 2 for method 2. There were $r = 10$ sets of whole plots. The color by label interaction is completely confounded with method in the SPED experiment performed.

Example 12—Three types of managements (factor A) constituted the whole plots that consisted of a litter of six male rats. The $b = 6$ medical treatments (factor B) were the split plot treatments with one rat constituting the speu. Three litters, wpeus, were obtained from each of $r = 6$ laboratories.

Example 13—A randomized complete block experiment design with $a = 5$ treatments (factor A) and $r = 5$ replicates was conducted to determine the effect of the treatments on the yield and the quality of strawberries. The experiment was laid out in the field in five columns, the blocks or replicates, and five rows. Hence, this is a row-column design as far as spatial variation is concerned. A 5×5 Latin square experiment design should have been used but was not. The strawberries in each of the 25 wpeus were graded into $b = 4$ quality grades (factor B) that were the split plot treatments. Responses were the weight and the number of strawberries in each of the grades within a wpeu.

Example 14—Jarmasz et al. (2005) used several forms of a split plot experiment design to study human subject perceptions to various stimuli. The factor sex was not taken into account when analyzing the data presented in the paper. Taking the factor sex into account adds to the splitting of units and the complexity of the analysis. Several variations of the SPED were used. The split-plot-designed experiment is of frequent occurrence in this type of research investigation.

Numerous literature citations of split plot designs are given by Federer (1955) in the Problem Section at the end of the book. This type of design appears in many fields of inquiry and is of frequent occurrence. Kirk (1968) lists ten references as representative applications of split plot designs in literature involving learning and other psychological research. The Annual Reports of the Rothamsted Experiment Station, the International Rice Research Institute (IRRI), and other research organizations give data sets for split-plot-designed experiments.

1.4. ANALYSIS OF VARIANCE

A partitioning of the degrees of freedom in an analysis of variance table for the various sources of variation is one method for writing a linear model for a set of experimental data. Alternatively, writing a linear model in equation form is another way of presenting the sources of variation for an experiment. A linear response model for the SPED for fixed effects factors A and B is usually given as

$$Y_{hij} = \mu + \rho_h + \alpha_i + \delta_{hi} + \beta_j + \alpha\beta_{ij} + \varepsilon_{hij}, \qquad (1.1)$$

where Y_{hij} is the response of the hijth speu,

μ is a general mean effect,

ρ_h is the hth replicate effect which is identically and independently distributed with mean zero and variance σ_ρ^2,

α_i is the effect of the ith whole plot factor A treatment,

δ_{hi} is a whole plot random error term which is identically and independently distributed with mean zero and variance σ_δ^2,

β_j is the effect of the jth split plot factor B treatment,

$\alpha\beta_{ij}$ is the interaction effect of the ith whole plot treatment with the jth split plot treatment, and

ε_{hij} is a split plot random error effect identically and independently distributed with mean zero and variance σ_ε^2.

The ρ_h, ε_{hi}, and δ_{hij} in Equation (1.1) are considered to be mutually independent variables.

Prior to calculating an analysis of variance, ANOVA table for the above response model, it is often instructive and enlightening to construct an ANOVA table for *each* whole plot as follows:

Whole plot level	A1		A2		...	Aa	
Source of variation	DF	SS	DF	SS	...	DF	SS
Total	rb	T1	rb	T2	...	rb	Ta
Correction for mean	1	C1	1	C2	...	1	Ca
Replicate	$r-1$	R1	$r-1$	R2	...	$r-1$	Ra
Split plot factor B	$b-1$	B1	$b-1$	B2	...	$b-1$	Ba
$R \times B =$ Error	$(r-1)(b-1)$	E1	$(r-1)(b-1)$	E2	...	$(r-1)(b-1)$	Ea

DF is degrees of freedom and SS is sum of squares. The dot notation is used which indicates that this is a sum over the subscripts replaced by a dot. The sums of squares for the ith whole plot treatment, $i = 1, 2, \ldots, a$, are:

$$Ti = \sum_{h=1}^{r} \sum_{j=1}^{b} Y_{hij}^2$$

$$Ci = Y_{.i.}^2/br$$

$$Ri = \sum_{h=1}^{r} Y_{hi.}^2/b - Y_{.i.}^2/br = b\sum_{i=1}^{r}(\bar{y}_{hi.} - \bar{y}_{.i.})^2$$

$$Bi = \sum_{j=1}^{b} Y_{.ij}^2/r - Y_{.i.}^2/br = r\sum_{j=1}^{b}(\bar{y}_{.ij} - \bar{y}_{.i.})^2.$$

These are the usual equations for computing sums of squares for data from a randomized complete block designed experiment. Ei is obtained by subtraction.

The estimated standard error of a difference between two factor B effects or means for random factor A effects is

$$\text{SE}(\bar{y}_{.j} - \bar{y}_{.j'}) = \sqrt{\frac{2\ A \times B \text{ interaction mean square}}{ra}}. \tag{1.5}$$

The estimated standard error of difference between two factor B effects or means at one level of factor A is

$$\text{SE}(\bar{y}_{.ij} - \bar{y}_{.ij'}) = \sqrt{\frac{2\ \text{Error } B \text{ mean square}}{r}}. \tag{1.6}$$

This latter standard error of a difference may be seen from the ANOVAs presented for each whole plot treatment. The estimated standard error of a difference between two factor A effects or means at one level of factor B is

$$\text{SE}(\bar{y}_{.ij} - \bar{y}_{.i'j}) = \sqrt{\frac{2[(b-1)\ \text{Error } B + \text{Error } A]}{rb}}. \tag{1.7}$$

The degrees of freedom for the above standard error of a difference and the following are unknown and need to be approximated.

The standard error of a mean for a whole plot treatment with random replicate effects is

$$\text{SE}(\bar{y}_{i.}) = \sqrt{\frac{\text{Error } A + b\sigma_\rho^2}{rb}}. \tag{1.8}$$

The standard error of a mean for a split plot treatment with random replicate effects is

$$\text{SE}(\bar{y}_{.j}) = \sqrt{\frac{\text{Error } B + a\sigma_\rho^2}{ra}}. \tag{1.9}$$

The standard error of an $A \times B$ interaction mean with random replicate effects is

$$\text{SE}(\bar{y}_{.ij}) = \sqrt{\frac{\text{Error } B + \sigma_\rho^2 + \sigma_\delta^2}{r}}. \tag{1.10}$$

The estimated values for the above standard errors of a mean and for difference between two means, Equations (1.3)–(1.10), are obtained by substituting the numerical values for Error A, Error B, and the estimate of the pertinent variance component for the corresponding ones in the above equations. The estimated values are obtained from an analysis of the data from an experiment.

1.7. NUMERICAL EXAMPLES

Example 1.1. A maize yield trial was conducted to determine the effects of four methods, $a = 4$, of primary seedbed preparations (A1, A2, A3, and A4), factor A the whole plot treatments, and four methods, $b = 4$, of planting the corn kernels (B1, B2, B3, and B4), factor B the split plot treatments. The basic split plot experiment design contained $r = 4$ complete blocks or replicates. The four seedbed preparations were arranged in a random fashion within each of the four replicates. Then within each of the $4 \times 4 = 16$ seedbed preparations, the wpeu was divided into four areas or plots, speus, and the four methods of planting maize seeds were randomly assigned to the four speus. The object of this experiment was to compare seedbed preparations and planting methods. In addition, it was desirable to know if there was an interaction and whether it is necessary to use a particular planting method for each seedbed preparation. A systematized arrangement of the maize yields from the experiment in bushels per acre, are given in Table 1.1.

Table 1.1. Bushels per Acre Yield of Maize for Seedbed Preparations and Planting Methods.

Replicate	Planting methods				Total
	B1	B2	B3	B4	
	A1 = plowed at 7 inches				
1	82.8	46.2	78.6	77.7	285.3
2	72.2	51.6	70.9	73.6	268.3
3	72.9	53.6	69.8	70.3	266.6
4	74.6	57.0	69.6	72.3	273.5
Total	302.5	208.4	288.9	293.9	1093.7
	A2 = plowed at 4 inches				
1	74.1	49.1	72.0	66.1	261.3
2	76.2	53.8	71.8	65.5	267.3
3	71.1	43.7	67.6	66.2	248.6
4	67.8	58.8	60.6	60.6	247.8
Total	289.2	205.4	272.0	258.4	1025.0
	A3 = blank basin listed				
1	68.4	54.5	72.0	70.6	265.5
2	68.2	47.6	76.7	75.4	267.9
3	67.1	46.4	70.7	66.2	250.4
4	65.6	53.3	65.6	69.2	253.7
Total	269.3	201.8	285.0	281.4	1037.5
	A4 = disk-harrowed				
1	71.5	50.9	76.4	75.1	273.9
2	70.4	65.0	75.8	75.8	287.0
3	72.5	54.9	67.6	75.2	270.2
4	67.8	50.2	65.6	63.3	246.9
Total	282.2	221.0	285.4	289.4	1078.0
B total	1142.2	836.6	1131.3	1123.1	

NUMERICAL EXAMPLES

Table 1.2. Analysis of Variance and F-Statistics for the Data of Table 1.1.

Source of variation	DF	Sum of squares	Mean square	F	Prob>F
Total	64	285,505.47			
Correction for mean	1	279,991.26			
Replicate = R	3	223.81	74.60		
Factor A	3	194.56	64.85	3.69	0.06
A1 + A4 vs. A2 + A3	1	186.32		10.60	0.01
Rest	2	8.24		0.47	
Error $A = R \times A$	9	158.24	17.58		
Planting method = B	3	4107.38	1369.13	81.01	0.00
B2 vs. rest	1	4105.15		242.90	0.00
Rest	2	2.23		0.13	
$A \times B$ interaction	9	221.74	24.64	1.46	0.20
Error B	36	608.48	16.90		

The grand total is 4234.2. The replicate totals are 1086.0, 1090.5, 1035.8, and 1021.9 for replicates 1, 2, 3, and 4, respectively. An analysis of variance and F-statistics for this experiment are given in Table 1.2. An SAS computer program for computing this analysis of variance table is given in Appendix 1.1.

The sum of squares for the contrast A1 + A4 − A2 − A3 is computed as

$$\frac{(1093.7 + 1078.0 - 1025.0 - 1037.5)^2}{64} = 186.32.$$

The sum of squares for the contrast 3(B2) − B1 − B3 − B4 is computed as

$$\frac{[3(836.6) - 1143.2 - 1131.3 - 1123.1]^2}{16(3^2 + 1 + 1 + 1)} = \frac{(-887.8)^2}{192} = 4105.15.$$

As may be observed, these two contrasts account for most of the differences among the planting methods and seedbed preparations. There is a slight indication that some interaction may be present. Also, since the two "Rest" mean squares are less than the Error A and Error B mean squares, there appears to be some type of heterogeneity that is not controlled. The problem of finding it is left as an exercise for the reader as is the computation for the interaction of the above two contrasts. Figures 1.1 and 1.2 illustrate the variation of planting methods in each of the seedbed preparations with two different axes.

A computer code for obtaining many of the numerical results including the means is given in Appendix 1.1.

Example 1.2. An experiment consisting of 49 guayule genotypes as whole plot treatments was designed as a triple lattice incomplete block experiment design with $r = 6$ replicates (see Federer, 1946). The split plot treatment represented four seed treatments for breaking the dormancy of guayule seeds. The split plot experimental unit consisted of 100 seeds planted in one-fourth of greenhouse flat. The wpeu was a greenhouse flat. Eight of the guayule genotypes from three of the six replicates from

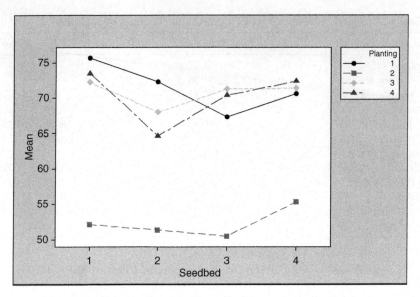

Figure 1.1. Planting method by seedbed preparation interaction.

this experiment were selected to illustrate the analysis for a split-plot-designed example. The selected data are analyzed as if a randomized complete block design had been used for the eight whole plot treatments. This design is now considered to be a standard split-plot-designed experiment. The data for the *ij* combinations of

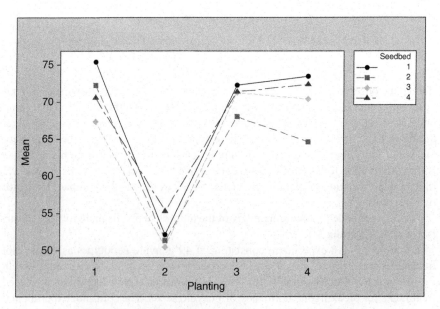

Figure 1.2. Planting method by seedling preparation interaction.

NUMERICAL EXAMPLES 19

genotypes $i = 0, 1, \ldots, 7$ and seed treatments $j = 0, 1, 2, 3$, for each of the replicates $h = 1, 2, 3$ are given in Table 1.3. The top number of a pair is the combination ij and the bottom number is the number of plants that emerged from 100 seeds. A computer program for computing the analysis of variance tables and means is given in Appendix 1.2. The genotype-by-seed treatment totals are given in Table 1.4. The ANOVAs for seed treatments by replicate for each genotype are presented in Table 1.5. An ANOVA for this split-plot-designed example is given in Table 1.6.

Table 1.3. Number of Plants Germinating from 100 Seeds for Each of Four Seed Treatments from Eight Guayule Genotypes in a Split Plot Experiment Design with Three Replicates.

	Replicate 1								
	01	23	30	52	42	11	73	61	
	12	10	52	28	9	26	9	12	
	02	20	33	53	43	12	71	62	
	13	51	13	14	12	27	14	26	
	00	21	32	51	40	10	72	63	
	66	8	19	8	45	77	30	15	
	03	22	31	50	41	13	70	60	
	6	20	4	59	20	15	49	56	
Total	97	89	88	109	86	145	102	109	825
	Replicate 2								
	32	60	73	41	03	12	51	21	
	16	38	15	13	12	5	8	16	
	31	62	70	43	00	10	52	22	
	15	16	41	12	63	47	32	30	
	33	61	72	40	02	11	53	20	
	9	16	28	51	13	11	21	81	
	30	63	71	42	01	13	50	23	
	40	8	20	10	10	4	66	14	
Total	80	78	104	86	98	67	127	141	781
	Replicate 3								
	63	72	50	42	32	22	01	11	
	7	36	49	12	7	29	13	18	
	62	71	52	40	30	21	03	10	
	24	25	29	52	59	14	7	66	
	61	70	53	41	31	23	00	12	
	16	54	16	16	11	10	70	11	
	00	73	51	43	33	20	02	13	
	45	12	8	11	7	63	11	15	
Total	92	127	102	91	84	116	101	110	823

Table 1.4. Genotype-by-Seed Treatment Totals.

Seed treatment	Genotypes								
	0	1	2	3	4	5	6	7	Total
0	199	190	195	151	148	174	139	144	1340
1	35	55	38	30	49	24	44	59	334
2	37	43	79	42	31	89	66	54	481
3	25	34	34	29	35	51	30	36	274
Genotype total	296	322	346	252	263	338	279	333	2449

From Table 1.5, we note that the residual mean squares vary from 8.417 for genotype 0 to 55.889 for genotype 3. This may indicate that a square root or arcsine transformation of the numbers is needed. This was not done as differences in seed treatment means are large and a more precise analysis may not be necessary. The F-values for seed treatments varied from 20.74 for genotype 3 to 276.12 for

Table 1.5. ANOVAs and F-Values for Replicate and Seed Treatment for Each Genotype, SS = Sum of squares, MS = Mean Square.

Genotype 0

Source	DF	SS	MS	F-value	Prob > F
Total	12	14,325.98	—	—	—
Correction for mean	1	7301.31	—	—	—
Replicate	2	2.167	1.08	0.13	0.88
Seed treatment	3	6972.00	2324.00	276.12	0.00
Residual	6	50.50	8.42	—	—

Genotype 1

Source	DF	SS	MS	F-value	Prob > F
Total	12	14,955.99	—	—	—
Correction for mean	1	8640.32	—	—	—
Replicate	2	763.17	381.58	15.31	0.00
Seed treatment	3	5403.00	1801.00	72.28	0.00
Residual	6	149.50	24.92	—	—

Genotype 2

Source	DF	SS	MS	F-value	Prob > F
Total	12	16,184.00	—	—	—
Correction for mean	1	9976.33	—	—	—
Replicate	2	338.17	169.08	4.53	0.06
Seed treatment	3	5645.67	1881.89	50.45	0.00
Residual	6	223.83	37.31	—	—

NUMERICAL EXAMPLES

Table 1.5. (*Continued*)

Genotype 3

Source	DF	SS	MS	F-value	Prob > F
Total	12	9112.00	—	—	—
Correction for mean	1	5292.00	—	—	—
Replicate	2	8.000	4.00	0.07	0.93
Seed treatment	3	3476.67	1158.89	20.74	0.00
Residual	6	335.33	55.89	—	—

Genotype 4

Source	DF	SS	MS	F-value	Prob > F
Total	12	8889.00	—	—	—
Correction for mean	1	5764.09	—	—	—
Replicate	2	4.17	2.08	0.23	0.80
Seed treatment	3	3066.25	1022.08	112.52	0.00
Residual	6	54.50	9.08	—	—

Genotype 5

Source	DF	SS	MS	F-value	Prob > F
Total	12	13,972.00	—	—	—
Correction for mean	1	9520.34	—	—	—
Replicate	2	83.167	41.58	2.56	0.16
Seed treatment	3	4271.00	1423.67	87.61	0.00
Residual	6	97.50	16.25	—	—

Genotype 6

Source	DF	SS	MS	F-value	Prob > F
Total	12	9107.00	—	—	—
Correction for mean	1	6486.75	—	—	—
Replicate	2	120.50	60.25	2.43	0.17
Seed treatment	3	2350.92	783.64	31.59	0.00
Residual	6	148.83	24.81	—	—

Genotype 7

Source	DF	SS	MS	F-value	Prob > F
Total	12	11,649	—	—	—
Correction for mean	1	9240.75	—	—	—
Replicate	2	96.50	48.25	2.82	0.14
Seed treatment	3	2208.92	736.31	42.96	0.00
Residual	6	102.83	17.14	—	—

Table 1.6. An Analysis of Variance and F-Values for the Data of Table 1.3.

Source of variation	DF	Sum of squares	Mean square	F-value	Prob > F
Total	96	98195	—	—	—
Correction for mean	1	61,458.76	—	—	—
Replicate	2	38.58	19.29	0.20	—
Genotype = A	7	763.16	109.02	1.11	—
Error A	14	1377.25	98.38	—	
Seed treatment = B	3	30,774.28	10,258.09	82.22	0.00
0 vs. 1 + 2 + 3	1	29,829.03		239.07	0.00
1 vs. 2 + 3	1	52.56		0.42	
2 vs. 3	1	892.69		7.15	0.02
Seed treatment × genotype	21	2620.13	124.77	5.15	0.00
A × 0 vs 1 + 2 + 3	7	1456.72	208.10	8.59	0.00
A × 1 vs. 2 + 3	7	632.94	90.42	3.73	0.01
A × 2 vs. 3	7	530.48	75.78	3.13	0.01
Error B	48	1162.84	24.23	—	—

genotype 0. There was smaller variation among replicate means, as the F-values varied from 0.07 for genotype 3 to 4.53 for genotype 2.

From the combined analysis in Table 1.6, it is noted that the seed treatment by genotype interaction is present. To study this in further detail, Figure 1.3 was prepared. Genotypes 0, 1, 2, and 5 for seed treatment 0 and genotypes 2 and 5 for

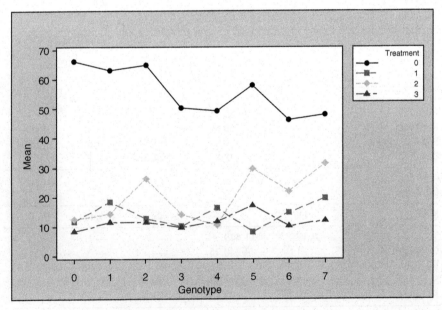

Figure 1.3. Genotype by seed treatment interaction.

seed treatment 2 are the ones contributing most to the interaction sum of squares. Their responses to seed treatments 0 and 2 were relatively higher than the other responses.

The various standard errors of a difference of two means are obtained next. The standard error of a difference between two genotype means is [Equation (1.3)].

$$\text{SE}(\bar{y}_{i.} - \bar{y}_{i'.}) = \sqrt{\frac{2(98.38)}{3(4)}} = \sqrt{16.3967} = 4.05.$$

The standard error of a difference of two seed treatment means for random genotype effects is [Equation (1.4)].

$$\text{SE}(\bar{y}_{.j} - \bar{y}_{.j'}) = \sqrt{\frac{2(124.77)}{3(8)}} = \sqrt{10.3975} = 3.22.$$

The standard error of difference between two seed treatment means for one genotype is [Equation (1.6)].

$$\text{SE}(\bar{y}_{.ij} - \bar{y}_{.ij'}) = \sqrt{\frac{2(24.23)}{3}} = 4.02.$$

The standard error of difference between two genotype means for one seed treatment is [Equation (1.7)].

$$\text{SE}(\bar{y}_{.ij} - \bar{y}_{.i'j}) = \sqrt{\frac{2\{3(24.23) + 98.38\}}{3(4)}} = \sqrt{28.5117} = 5.34.$$

Even though a seed treatment by genotype interaction is present (see Figure 1.3.), the mean for seed treatment 0, threshed seed and untreated, is far above the remaining three seed treatment means in increasing the germination percentage of guayule seeds, hence would be preferred for treating guayule seeds. Seed treatment 1 was unthreshed and untreated, seed treatment 2 was unthreshed and treated 1943 seeds, and seed treatment 3 was unthreshed and treated 1942 seeds. The seed treatment 3 represented the seed treatment method prior to this experiment. Seed treatment 0 involved using threshed and untreated seeds and was quite effective in increasing seed germination. The threshing removed many empty seeds. There are also genetic differences in a genotype response to seed treatment 0 but in all cases this treatment was superior to the other treatments. Some genotypes may have more empty seeds than others. A further experiment using different lengths of threshing times or other treatments may be needed when specific genotypes are being considered.

1.8. MULTIPLE COMPARISONS OF MEANS

Federer and McCulloch (1984) presented multiple comparison procedures for an SPED. They discussed five different multiple comparison procedures, viz., the least significant difference procedure (lsd) with a per comparison error rate, Tukey's range procedure or honestly significant difference (hsd) with an experiment-wise error rate for all pairs of two means, the Bonferroni procedure (esd) with a per experiment error rate for m specified contrasts, Scheffe's procedure (ssd) with an error rate for all possible comparisons and contrasts, and Dunnett's procedure (dsd) for comparing treatments with a control with an experiment-wise error rate.

Using the data given in Table 1.3, all possible differences between pairs of the eight guayule genotype (whole plot) means are given in Table 1.7.

All possible differences between pairs of the seed treatment (split plot) means are presented in Table 1.8.

A two-sided Type I error rate of 5% is used. For genotype mean differences, the lsd is computed as (df = degrees of freedom associated with error term and E_a is the Error A mean square),

$$t_{.05,14\text{df}}\sqrt{\frac{2E_a}{rb}} = 2.145\sqrt{\frac{2(98.38)}{3(4)}} = 8.69. \tag{1.11}$$

Using Tukey's procedure, the hsd is computed as,

$$q_{.05,8,14\text{df}}\sqrt{\frac{E_a}{rb}} = 4.99\sqrt{\frac{98.38}{3(4)}} = 14.29. \tag{1.12}$$

Using the Bonferroni procedure, the esd, for $m = \dfrac{8(8-1)}{2} = 28$, is computed as,

$$\text{esd} = t_{.05/m,14\text{df}}\sqrt{\frac{2E_a}{rb}} = 3.85\sqrt{\frac{2(98.38)}{3(4)}} = 15.57. \tag{1.13}$$

Table 1.7. All Possible Differences Between Pairs of Eight Guayule Genotype Means, Number of Seeds Germinated Out of 100.

		Genotype						
Genotype	Mean	2 29	5 28	7 28	1 27	0 25	6 23	4 22
3	21	8	7	7	6	4	2	1
4	22	7	6	6	5	3	1	—
6	23	6	5	5	4	2	—	—
0	25	4	3	3	2	—	—	—
1	27	2	1	1	—	—	—	—
7	28	1	0	—	—	—	—	—
5	28	1	—	—	—	—	—	—

MULTIPLE COMPARISONS OF MEANS

Table 1.8. All Possible Differences Between Pairs of the four
Seed Treatment Means, Number of Seeds Germinated out of 100.

		Seed treatment		
		0	2	1
Seed treatment	Mean	56	20	14
3	11	45	9	3
1	14	42	6	—
2	20	36	—	—

Using Scheffe's method, the ssd is computed as (df = degrees of freedom for Error A),

$$\text{ssd} = \sqrt{\frac{2(a-1)F_{.05}(a-1,\text{df})(E_a)}{rb}} = \sqrt{7(2.77)(2)\left\{\frac{98.38}{3(4)}\right\}} = 17.83. \quad (1.14)$$

Using the Dunnett method, the dsd is computed with genotype 2, for example, designated as the control treatment,

$$\text{dsd} = d_{a-1,\text{df},.05}\sqrt{\frac{2E_a}{rb}} = 3.10(4.045) = 12.54. \quad (1.15)$$

To compare differences between seed treatment means for random genotype effects, the lsd is computed as (E_{ab} is the interaction mean square),

$$\text{lsd} = t_{.05,21\text{df}}\sqrt{\frac{2E_{ab}}{ra}} = 2.08\sqrt{\frac{2(124.77)}{3(8)}} = 6.71. \quad (1.16)$$

The hsd is computed as,

$$q_{.05,4,21}\sqrt{\frac{E_{ab}}{ra}} = 3.96\sqrt{\frac{124.77}{3(8)}} = 9.03. \quad (1.17)$$

The esd, for $m = 4(4-1)/2 = 6$, is computed as,

$$\text{esd} = t_{.05/m,21\text{df}}\sqrt{\frac{2(E_{ab})}{ra}} = 3.82\sqrt{\frac{2(124.77)}{3(8)}} = 9.13. \quad (1.18)$$

The ssd is computed as (df = degrees of freedom for error mean square for seed treatments),

$$\text{ssd} = \sqrt{\frac{2(b-1)F_{.05}(b-1,\text{df})(E_{ab})}{ra}} = \sqrt{3(3.07)(2)\left\{\frac{124.77}{3(8)}\right\}} = 9.79. \quad (1.19)$$

The dsd is computed as, where seed treatment 2 is designated as the control treatment,

$$\text{dsd} = d_{b-1,\text{df},.05}\sqrt{\frac{2(E_{ab})}{rb}} = 2.56\sqrt{\frac{2(124.77)}{3(8)}} = 2.56(3.225) = 8.25. \quad (1.20)$$

In order to obtain the values for hsd, esd, and dsd for large values, say $ab = 32$, more extensive tables of t, q, and d values are needed. For the 32 genotype by seed treatment mean differences, the lsd for comparing seed treatment means for a specific genotype is computed as

$$\text{lsd} = t_{.05,48\text{df}}\sqrt{\frac{2E_b}{r}} = 2.01\sqrt{\frac{2(24.23)}{3}} = 2.01(4.019) = 8.08. \quad (1.21)$$

The ssd is computed as,

$$\text{ssd} = \sqrt{\frac{2(ab-1)F_{.05}(ab-1,\text{df})(E_b)}{r}} = \sqrt{2(31)(1.76)\left\{\frac{24.23}{3}\right\}} = 29.69. \quad (1.22)$$

The above multiple comparison results were for the genotype and seed treatment means. If it is desirable to perform multiple comparisons on the 32 genotypes by seed treatment means, the standard error of means and of differences between two means given in Equations (1.3)–(1.10) will need to be used. Depending upon which pair of means and method is under consideration, the appropriate standard error of a mean or of a difference between two means will need to be selected for each pair.

1.9. ONE REPLICATE OF A SPLIT PLOT EXPERIMENT DESIGN AND MISSING OBSERVATIONS

In the course of statistical consulting, many types of experiment designs are encountered. Federer (1975) presents an example wherein only one replicate of a split plot experiment design was used, and the experimenter wanted advice on the statistical analysis. Three light intensities were the whole plot treatments and were used in three different growth chambers. The three plant types were the split plot treatments. The plant types were tomato, pigweed, and pigweed + tomato (an intercropping treatment as described by Federer, 1993, 1999). One greenhouse flat of eight plants formed the speu. Eight tomato plants and eight pigweed plants were used for the tomato and pigweed plant types. Four pigweed and four tomato plants made up the tomato + pigweed treatment. An ANOVA for one light intensity in one growth chamber is as given below in the tabular form.

Source of variation	DF	DF general
Total	24	bk
Correction for mean	1	1
Replicates (blocks) = R	0	0
Plant types = B	2	$b - 1$
$R \times B$	0	0
Plants within B	21	$b(k - 1)$

The $R \times B$ error mean square is the appropriate error term for comparing plant-type means but there is no estimate of it. The plants within B mean square are frequently used in place of the $R \times B$ interaction mean square. This is incorrect in that the variance component due to variation from block to block is not included in the plants within B mean square but it is in the plant-type mean square. This is a frequent mistake found in published literature. An ANOVA partitioning of the degrees of freedom for all 72 observations is as given below in the tabular form.

Source of variation	DF	DF general
Total	72	abk
Correction for mean	1	1
Replicates (blocks) = R	0	0
Light intensity = chamber = A	2	$a - 1$
Error $A = R \times A$	0	0
Plant types = B	2	$b - 1$
$A \times B$	4	$(a - 1)(b - 1)$
Error $B = R \times B$ within A	0	0
Plants within $R, A,$ and B	63	$ab(k - 1)$

If light intensity could be considered to be a random effect, then the $A \times B$ mean square would be used as the error term for comparing the plant-type means but it has only four degrees of freedom. The fact that there are $63 = ab(k - 1)$ degrees of freedom associated with the plants within $R, A,$ and B mean squares, has tempted many researchers to use these mean squares to replace the appropriate Error B mean square. This practice results in using an error mean square that is too small, resulting in false significance statements.

Occasionally missing observations occur in split-plot-designed experiments. Anderson (1946) and Khargonkar (1948) present formulas for computing missing plot values. Computer packages are mostly designed to handle these situations and hence there is no need for the formulas.

1.10. NATURE OF EXPERIMENTAL VARIATION

In Section 1.5, several assumptions about the nature of the random error terms were made to obtain the expected values of the various mean squares. The split plot random errors ε_{hij} were assumed to be identically and independently distributed, IID$(0, \sigma_\varepsilon^2)$. It was further assumed that the Error A mean square contained both terms σ_ε^2 and σ_δ^2 with the split plot, whole plot random errors being additives. Also, it was assumed that the split plot and whole plot random error terms were independent. Simply because this was assumed and because it appears to be a reasonable assumption, does not mean that the assumption holds for all split-plot-designed experiments.

Cochran and Cox (1957, Section 7.12) and Kirk (1968, Chapter 8) present another way of quantifying the experimental variation exhibited by a split-plot-designed experiment. These authors assume that the split plot random errors ε_{hij} are correlated. For spatially laid out experiments, this correlation may be due to the proximity of neighboring speus. In baking and industrial experiments, a single batch may be divided into b speus for the b split plot treatments. Any factor affecting the batch affects all b speus. They assumed that the following correlation structure holds:

$$E[\varepsilon_{hik}\varepsilon_{hij}] = \rho\sigma_\varepsilon^2 \text{ and } E[\varepsilon_{hij}\varepsilon_{rst}] = 0, \ j \neq k, \ hij \neq rst \qquad (1.23)$$

The random split plot error terms in the same whole plot have covariance $\rho\sigma_\varepsilon^2$ and those in different whole plots have zero covariance, that is, they are un-correlated.

Consider the case where factor B has $b = 2$ levels. Ignoring the random nature of the replicate or complete block effect, the variance of a whole plot is

$$E[(\varepsilon_{hi1} + \varepsilon_{hi2})^2] = \sigma_\varepsilon^2 + \sigma_\varepsilon^2 + 2\rho\sigma_\varepsilon^2 = 2\sigma_\varepsilon^2(1 + \rho). \qquad (1.24)$$

For b levels of split plot treatments, the whole plot variance is,

$$E[(\varepsilon_{hi1} + \varepsilon_{hi2} + \varepsilon_{hi3} + \cdots + \varepsilon_{hib})^2] = b\sigma_\varepsilon^2(1 + (b-1)\rho). \qquad (1.25)$$

The split plot main effects are derived from differences of split plot responses. Hence the variance of a difference of two split plot treatments is,

$$E[(\varepsilon_{hi1} - \varepsilon_{hi2})^2] = \sigma_\varepsilon^2 + \sigma_\varepsilon^2 - 2\rho\sigma_\varepsilon^2 = 2\sigma_\varepsilon^2(1 - \rho), \qquad (1.26)$$

with an effective error variance per speu of $\sigma_\varepsilon^2(1 - \rho)$ whatever the value of b. This variance also applies to contrasts of interaction effects within the same whole plot. For comparing two interaction terms from different whole plots the variance is $2\sigma_\varepsilon^2$. The analysis of variance, as described above, gives unbiased and correct estimates of the above.

For many situations ρ will be positive. However, it could be negative for certain types of experimental variations and materials. In this event, the Error A mean square could be smaller than the Error B mean square. Also, if competition is present among the speus within the same whole plot, the Error B mean square would contain a component of variance due to competition that would not enter into the Error A mean square. This could make Error B larger than the Error A mean square. Another situation where this could occur is when there is more genetic variation within whole plots than among whole plots. It is also possible owing to lack of symmetry of the distribution of random split plot effects that a transformation of the responses may be required in order to have Error B less than Error A. Numerical examples can easily be constructed where Error B is considerably larger than Error A.

1.11. REPEATED MEASURES EXPERIMENTS

Some authors, for example, Kirk (1968, Chapter 8), consider a repeated measures experiment as a split-plot-designed experiment. There are two kinds of repeated measures experiments, viz., the same treatment is repeated b times on a single subject or the b treatments of factor B are applied sequentially to a subject over b periods. The latter type of experiment is known as a cross-over-designed experiment. It is this type of repeated measures experiment that has been confused with split-plot-designed experiments. It is inappropriate to consider the cross-over-designed experiment as a split plot experiment. One reason is that there are several kinds of treatment effects in a cross-over experiment, whereas, there is only one kind of treatment effect for split plot experiments. A cross-over experiment may have the direct effect of a treatment in the period in which it was applied, a carryover effect in the periods after it has been applied, a continuing effect, and/or a permanent effect. A split plot experiment has only the direct effect of the treatment. Also the treatment design is different for a cross-over experiment as certain sequences of treatments on a subject are used for a cross-over design, whereas the treatments in a split plot design appear in a random order. The complexity of the nature of treatment effects and the statistical design in a cross-over experiment makes it prudent to consider this class of designs as an entity in itself. Therefore, this type of designed experiment should not be confused with split-plot-designed experiments.

1.12. PRECISION OF CONTRASTS

The average overall precision of the contrasts in a standard split plot design is the same as that for a randomized complete block design of the ab treatment combinations. The precision of whole plot treatment contrasts, factor A, is usually less than or equal to what it would be for a randomized complete block design. The gain in precision is obtained for the split plot treatments, factor B, and for the interaction effects. Thus, if less precision is required for factor A treatments and more for factor B and interaction effects, the split plot design is admirably suited for this situation. Another reason that a split plot design may be selected is that larger experimental units are required for factor A treatments than for factor B treatments. For example, fertilizer and irrigation treatments require larger experimental units than do treatments like varieties, pesticides, etc.

Federer (1955, page 274) presents a measure of the efficiency for split plot treatments (Also, see Kempthorne, 1952, Section 19.4). Using this measure, the precision of the split plot treatments, factor B, and of the $A \times B$ interactions is

$$\frac{(a-1)(\sigma_\varepsilon^2 + b\sigma_\delta^2) + a(b-1)\sigma_\varepsilon^2}{(ab-1)\sigma_\varepsilon^2} = 1 + \frac{b(a-1)\sigma_\delta^2}{(ab-1)\sigma_\varepsilon^2}, \qquad (1.27)$$

and the precision is estimated by,

$$\frac{(a-1)\text{Error } A + a(b-1)\text{Error } B}{(ab-1)\text{Error } B}. \qquad (1.28)$$

The precision of the whole plot treatment effects, factor A, is given by,

$$\frac{(a-1)(\sigma_\varepsilon^2 + b\sigma_\delta^2) + a(b-1)\sigma_\varepsilon^2}{(ab-1)(\sigma_\varepsilon^2 + b\sigma_\delta^2)} = \frac{(a-1)b\sigma_\delta^2 + (ab-1)\sigma_\varepsilon^2}{(ab-1)(\sigma_\varepsilon^2 + b\sigma_\delta^2)}, \qquad (1.29)$$

and it is estimated by,

$$\frac{(a-1)\text{Error } A + a(b-1)\text{Error } B}{(ab-1)\text{Error } A}. \qquad (1.30)$$

If the variance component σ_δ^2 is equal to zero, the precision is equal to one in both cases.

For the numerical example in Section 1.7, Example 1.2 with $a = 8$ and $b = 4$, the estimated precision for the split plot treatments, seed treatments, and interaction from Equation (1.28) is,

$$\frac{(8-1)(98.38) + 8(4-1)(24.23)}{\{8(4)-1\}(24.23)} = 1.69,$$

that is, a 69% increase over conducting the experiment as a randomized complete block design. The estimated precision for the whole plot treatments, guayule genotypes, from equation (1.30) is,

$$\frac{(8-1)(98.38) + 8(4-1)(24.23)}{\{8(4)-1\}(98.38)} = 0.42,$$

that is, a 58% loss over using a randomized complete block experiment design.

1.13. PROBLEMS

Problem 1.1. For the data of Example 1.1

(i) Analyze the data and use residual diagnostic plots to assess the equal variance and normality assumptions.

(ii) Use a multiple comparison procedure to test significance of pairs of means.

(iii) Redo (i) and (ii) for log transformed data.

Problem 1.2. For the data of Example 1.2

(i) Using the square root, transform the data, and perform analysis.

(ii) Use the arcsine transformation and obtain an analysis of the data.

(iii) Are there any differences in interpretation from these analyses?

Problem 1.3. Mazur (2005) presents several data sets for split-plot-designed experiments. For one of the experiments, four rats, R1, R2, R3, and R4, represented the four blocks or replicates, two whole plot treatments representing two time levels, long (20 s) and short (10 s), and four time intervals between stimuli, B1, B2, B3, and B4, were the split plot treatments. The terminal link entries per hour to the stimuli were:

| | R1 | | R2 | | R3 | | R4 | |
Condition	Long	Short	Long	Short	Long	Short	Long	Short
B1 = 60s	42.8	48.5	51.7	60.0	45.7	49.3	56.8	59.7
B2 = 30s	108.8	53.9	97.9	51.9	86.9	50.8	103.9	56.8
B3 = 15s	100.4	43.7	198.3	54.4	161.4	53.5	211.1	55.7
B4 = 2s	59.8	51.3	899.1	1.9	176.1	51.6	614.9	50.1

(i) Give a linear model for an analysis of these data and state assumptions used.

(ii) Obtain an analysis for these data and interpret the results.

(iii) Use residual diagnostic plots to assess the equal variance and normality assumptions.

(iv) If the assumptions are violated, find a transformation of the data that is suitable to obtain more variance homogeneity.

Problem 1.4. The data for this problem were taken from Mazur (2005). A split plot designed experiment on four pigeons, P1, P2, P3, and P4, as the four blocks was used. Two delay intervals, long (20 s) and short (10 s) represented the whole plot treatments. The eight split plot treatments, B1, B2, B3, B4, B5, B6, B7, and B8, were in a two by four factorial treatment arrangement. The two levels of one factor, say F, were independent, ind, and dependent, dep, and the four levels of the second factor were variations of presenting the four levels of the second factor, T. The levels of T are those described in Problem 1.3. The terminal entry rates per hour are given in the table below:

Condition	P1		P2		P3		P4	
	Long	Short	Long	Short	Long	Short	Long	Short
B1 = 60sind	31.5	62.3	42.7	52.9	40.4	61.6	5.9	61.8
B2 = 30sind	14.7	67.3	42.1	57.3	19.3	61.2	5.6	57.1
B3 = 15sind	12.2	63.1	42.1	61.2	22.0	56.5	0.8	57.4
B4 = 2sind	8.1	54.8	43.0	60.2	42.0	54.2	8.9	61.1
B5 = 60sdep	38.2	40.0	45.7	46.2	37.6	37.6	25.7	25.3
B6 = 30sdep	62.3	30.4	77.1	39.0	59.1	29.4	31.7	16.4
B7 = 15sdep	104.2	25.2	140.9	34.8	88.6	21.9	28.6	7.5
B8 = 2sdep	124.3	4.1	339.0	11.7	178.8	6.6	69.8	2.4

  (i) Obtain an analysis of the data.
  (ii) Is variance heterogeneity a problem?
  (iii) Obtain an analysis omitting P4 data. Does this change the results?
  (iv) Discuss the results of your various analyses and compare the results on rats in the previous problem with those on pigeons.

Problem 1.5. For the data given as Example 2 in Section 2

  (i) Write a SAS/PROC GLM code for obtaining an analysis of variance and means.
  (ii) Prepare a graph of the corn genotype by district interaction. Is there a significant district by corn genotype interaction?
  (iii) Compute the residuals and perform a diagnostic plot to assess the variance homogeneity and normality assumptions.

Problem 1.6. For the data of Example 4 in Section 2

  (i) Perform a multiple comparisons procedure using the lsd, the hsd, and the esd methods for the 10 combinations of alfalfa and no alfalfa with the five bromegrass strains. Are there significant differences among the pairs of means?
  (ii) Compute Tukey's one degree of freedom for non-additivity. Is there an indication of non-additivity?

1.14. REFERENCES

Anderson, R. L. (1946). Missing-plot techniques. Biometrics 2:41–47.

Cochran, W. G. and G. M. Cox. (1957). *Experimental Designs,* Second edition. John Wiley & Sons, Inc., New York, pp. 293–316.

Das, M. N. and N. C. Giri. (1979). *Design and Analysis of Experiments*, John Wiley & Sons, New York, Chichester, Brisbane, Toronto, pp. 149–151.

Federer, W. T. (1946). Variability of certain seed, seedling, and young-plant characters of guayule. *United States Department of Agriculture*, Technical bulletin No. 919, pp. 1–25.

Federer, W. T. (1955). *Experimental Design: Theory and Application*, Macmillan, New York (Reprinted in 1662 and 1974 by Oxford &IBH Publishing Company, New Delhi), Chapter X, pp. 271–306.

Federer, W. T. (1975). The misunderstood split plot. In *Applied Statistics* (Editor: R. P. Gupta), North-Holland Publishing Company, Amsterdam, Oxford, pp. 9–39.

Federer, W. T. (1984). Principles of statistical design with special reference to experiment and treatment design. In *Statistics: An Appraisal-Proceedings of the 50th Anniversary Conference, Iowa State Statistics Laboratory* (Editors: H. A. David and H. T. David), Iowa State University Press, Ames, Iowa, pp. 77–104.

Federer, W. T. (1991). *Statistics and Society: Data Collection and Interpretation*, Second edition. Marcel Dekker, Inc., New York, Chapter 7, pp. 1–578.

Federer, W. T. (1993). *Statistical Design and Analysis for Intercropping Experiments Volume I: Two Crops*, Springer-Verlag, New York, Berlin, Heidelberg, London, Paris, Tokyo, Hong Kong, Barcelona, Budapest, pp. 1–298.

Federer, W. T. (1999). *Statistical Design and Analysis for Intercropping Experiments Volume II: Three or More Crops*, Springer-Verlag, New York, Berlin, Heidelberg, London, Paris, Tokyo, Hong Kong, Barcelona, Budapest, pp. 1–262.

Federer, W. T. (2003). Exploratory model selection for spatially designed experiments. J Data Sci 1:231–248.

Federer, W. T. and C. E. McCulloch. (1984). Multiple comparisons procedures for some split plot and split block designs. In *Design of Experiments: Ranking and Selection* (*Essays in Honor of Robert E. Bechhofer*) (Editors: T. J. Santner and A. C. Tamhane), Marcel Dekker, Inc., New York, pp. 7–22.

Fisher, R. A. (1966). *The Design of Experiments*, Eighth edition. Oliver and Boyd, Edinburgh, (First edition, 1935), pp. 1–248.

Gomez, K. A. and A. A. Gomez. (1984). *Statistical Procedures for Agricultural Research*, John Wiley & Sons, New York, Chichester, Brisbane, Toronto, Singapore, pp. 101–110.

GENDEX (2005). http://designcomputing.net/gendex/modules.html.

Kempthorne, O. (1952). *The Design and Analysis of Experiments*, John Wiley & Sons, Inc., New York, Chapter 19, pp. 1–631.

Khargonkar, S. A. (1948). The estimation of missing value in split-plot and strip-plot trials. J Indian Soc of Agric Stat 1:147–161.

Jarmasz, J., C. M. Herdman, and K. R. Johannsdottier. (2005). Object based attention and cognitive tunneling. J Exp Psychol 11(1):3–12.

Kirk, R. E. (1968). *Experimental Design: Procedures for the Behavioral Sciences*, Brooks/Cole Publishing Company, Belmont, California, pp. 245–318.

Leonard, W. H. and A. G. Clark. (1938). *Field Plot Technique*, Class Notes, Agronomy Department, Colorado State University, Fort Collins, Colorado, copyright by authors.

Mazur, J. E. (2005). Exploring a concurrence-chains paradox decreasing preference as an initial link is shortened. J Exp Psychol: Anim Behav 31(1):3–17.

Mead, R. (1988). *The Design of Experiments: Statistical Principles for Practical Application*, Cambridge University Press, Cambridge, New York, Port Chester, Melbourne, Sydney, Chapter 14, pp. 1–628.

Raghavarao, D. (1983). *Statistical Techniques in Agricultural Research*, Oxford & IBH Publishing Company, New Delhi, Bombay, Calcutta, Section 11.15, pp. 1–271.

Snedecor, G. W. and W. G. Cochran. (1980). *Statistical Methods, Seventh edition*. Iowa State University Press, Ames, Iowa, pp. 1–507.

APPENDIX 1.1. EXAMPLE 1.1 CODE

The following is the SAS PROC/GLM code for obtaining an analysis of the data for Example 1.1:

```
Data spex1;

input Y R A B; /*Y=yield, R=block, A=planting method,
B=cultivation method*/

datalines;

82.8 1 1 1
46.2 1 1 2
78.6 1 1 3

...

65.6 4 4 3
63.3 4 4 4

;run;

Proc GLM;

Class R A B;
Model Y = R A R*A B A*B;
Lsmeans A B A*B;
Test H = A E = A*R;

Run;
```

The following is an abbreviated form of the output for the above code for Example 1.1:
Output
Dependent Variable: Y

APPENDIX

Source	DF	Sum of Squares	Mean Square	F Value	Pr > F
Model	27	4905.738750	181.694028	10.75	<.0001
Error	36	608.478750	16.902187		
Corrected Total	63	5514.217500			

R-Square	Coeff Var	Root MSE	Y Mean
0.889653	6.215594	4.111227	66.14375

Source	DF	Type I SS	Mean Square	F Value	Pr > F
R	3	223.808750	74.602917	4.41	0.0096
A	3	194.561250	64.853750	3.84	0.0176
R*A	9	158.242500	17.582500	1.04	0.4284
B	3	4107.383750	1369.127917	81.00	<.0001
A*B	9	221.742500	24.638056	1.46	0.2012

Source	DF	Type III SS	Mean Square	F Value	Pr > F
R	3	223.808750	74.602917	4.41	0.0096
A	3	194.561250	64.853750	3.84	0.0176
R*A	9	158.242500	17.582500	1.04	0.4284
B	3	4107.383750	1369.127917	81.00	<.0001
A*B	9	221.742500	24.638056	1.46	0.2012

Least Squares Means

A	Y LSMEAN
1	68.2937500
2	64.0625000
3	64.8437500
4	67.3750000

B	Y LSMEAN
1	71.3875000
2	52.2875000
3	70.7062500
4	70.1937500

A	B	Y LSMEAN
1	1	75.3750000
1	2	52.1000000
1	3	72.2250000
1	4	73.4750000
2	1	72.3000000
2	2	51.3500000
2	3	68.0000000
2	4	64.6000000

3	1	67.3250000
3	2	50.4500000
3	3	71.2500000
3	4	70.3500000
4	1	70.5500000
4	2	55.2500000
4	3	71.3500000
4	4	72.3500000

Dependent Variable: Y

Tests of Hypotheses Using the Type III MS for R*A as an Error Term

Source	DF	Type III SS	Mean Square	F Value	Pr > F
A	3	194.5612500	64.8537500	3.69	0.0557

APPENDIX 1.2. EXAMPLE 1.2 CODE

A SAS PROC/GLM code for obtaining an analysis for the data of Example 1.2 is given below. The code for an analysis for each whole plot data set is obtained by using IF and THEN statements such as "IF A > 1 THEN DELETE; and IF A < 1 THEN DELETE;" to obtain an analysis of the data for genotype 1. An analysis is obtained for the entire data set using the following code:

```
Data spex2;
Input Y R A B; /*Y=count, R=block, A=genotype, B=seed treatment*/
Datalines;
12  1  0  1
13  1  0  2
66  1  0  0
...
11  3  1  2
15  3  1  3
;
Proc GLM;
Class R A B ;
Model Y = R A R*A B A*B;
Lsmeans A B A*B;
Test H = A E = R*A;

run;
```

An abbreviated form of the output from running the above code is given below:

APPENDIX

Dependent Variable: Y

Source	DF	Sum of Squares	Mean Square	F Value	Pr > F
Model	47	35573.40625	756.88098	31.24	<.0001
Error	48	1162.83333	24.22569		
Corrected Total	95	36736.23958			

R-Square	Coeff Var	Root MSE	Y Mean
0.968346	19.45279	4.921960	25.30208

Source	DF	Type I SS	Mean Square	F Value	Pr > F
R	2	38.58333	19.29167	0.80	0.4568
A	7	763.15625	109.02232	4.50	0.0006
R*A	14	1377.25000	98.37500	4.06	0.0001
B	3	30774.28125	10258.09375	423.44	<.0001
A*B	21	2620.13542	124.76835	5.15	<.0001

Source	DF	Type III SS	Mean Square	F Value	Pr > F
R	2	38.58333	19.29167	0.80	0.4568
A	7	763.15625	109.02232	4.50	0.0006
R*A	14	1377.25000	98.37500	4.06	0.0001
B	3	30774.28125	10258.09375	423.44	<.0001
A*B	21	2620.13542	124.76835	5.15	<.0001

Least Squares Means

A	Y LSMEAN
0	24.6666667
1	26.8333333
2	28.8333333
3	21.0000000
4	21.9166667
5	28.1666667
6	23.2500000
7	27.7500000

B	Y LSMEAN
0	55.8333333
1	13.9166667
2	20.0416667
3	11.4166667

A	B	Y LSMEAN
0	0	66.3333333
0	1	11.6666667
0	2	12.3333333
0	3	8.3333333
1	0	63.3333333

1	1	18.3333333
1	2	14.3333333
1	3	11.3333333
2	0	65.0000000
2	1	12.6666667
2	2	26.3333333
2	3	11.3333333
3	0	50.3333333
3	1	10.0000000
3	2	14.0000000
3	3	9.6666667
4	0	49.3333333
4	1	16.3333333
4	2	10.3333333
4	3	11.6666667
5	0	58.0000000
5	1	8.0000000
5	2	29.6666667
5	3	17.0000000
6	0	46.3333333
6	1	14.6666667
6	2	22.0000000
6	3	10.0000000
7	0	48.0000000
7	1	19.6666667
7	2	31.3333333
7	3	12.0000000

CHAPTER 2

Standard Split Block Experiment Design

2.1. INTRODUCTION

A *split block experiment design* (Yates, 1933; Federer, 1955, Section X-2; Lentner and Bishop, 1986, Section 11.5) has also gone under different names including strip block design (e.g., Gomez and Gomez,1984), a two-way whole plot design, strip-plot (Nair, 1944; Khargonkar, 1948; Hoshmand, 1994, Section 5.6), sub-treatments in strips across blocks (Leonard and Clark, 1939; Cochran and Cox, 1950), and a criss-cross design (e.g., Mead, 1988). A factorial treatment design is usually involved. Suppose the two factors are A with a levels and B with b levels. Factor A (or B) may consist of a factorial or other treatment design. In each complete block (replicate), the a levels of factor A are randomly allotted to the a whole plot experimental units, *wpeusa*. Then perpendicular to the A *wpeusa*, the B experimental units, *wpeusb*, are formed and the b levels of factor B are randomly allocated to the second set of b whole plot units in each of the complete blocks. The levels of factor B go across all levels of factor A and likewise the levels of factor A go across all levels of factor B in a criss-cross manner. This arrangement with a different randomization is repeated in each of the r complete blocks. For this design, there are two whole plot treatments, A and B, and two whole plot experimental units, *wpeua* for factor A and *wpeub* for factor B. Note that the $A \times B$ interaction effects are in a split plot arrangement to both of the whole plot treatments. A randomized complete block experiment design is used for factor A treatments. Also, the factor B treatments are arranged in a randomized complete block experiment design. There are r randomizations for the levels of factor A and r randomizations for the levels of factor B. This is the design for a standard split block experiment design although any other design may be used for the whole plot treatments factors A and/or B.

Variations on Split Plot and Split Block Experiment Designs, by Walter T. Federer and Freedom King
Copyright © 2007 John Wiley & Sons, Inc.

A schematic layout of a split block experiment design consisting of a two-way array of factors A and B is given below:

	Block 1					Block 2					Block r			
A	1	2	...	a	1	2	...	a	1	2	...	a
B 1												
2												
3												
...												
b												

The following steps illustrate the randomization procedure and the layout of a split block design with $r = 4$ blocks, factor A with $a = 3$ levels in a randomized complete block design, and factor B with $b = 5$ levels in a randomized complete block design:

Step 1. Group similar experimental units into four blocks.

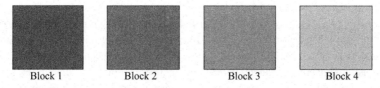

Step 2. Randomization of factor A levels to experimental units in each block.

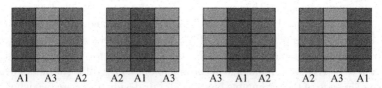

Step 3. Randomization of factor B levels to experimental units across factor A levels.

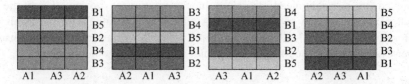

To show the effect of this on the statistical analysis of an experiment designed in this fashion, consider the following analysis of variance table, ANOVA, showing the partitioning of the degrees of freedom for the various sources of variation:

Source of variation	Degrees of freedom
Total	$rab = 4(3)(5) = 60$
Correction for mean	1
Replicate $= R$	$r - 1 = 3$
Factor A	$a - 1 = 2$
$R \times A =$ error A	$(r-1)(a-1) = 6$
Factor B	$b - 1 = 4$
$R \times B =$ error B	$(r-1)(b-1) = 12$
$A \times B$	$(a-1)(b-1) = 8$
$R \times A \times B =$ error AB	$(r-1)(a-1)(b-1) = 24$

Since the experimental units are different for factor A, for factor B, and for the $A \times B$ interaction, three different error terms are required in this design. There are r randomizations for each of the two whole plot treatments but since their experimental units, eus, are different, they have different error mean squares.

2.2. EXAMPLES

Example 1—Snedecor and Cochran (1980), section 16.15, present the data for an experiment involving three alfalfa genotypes arranged in a randomized complete block design with $r = 6$ replicates. This example brings up a number of considerations. If the analysis had been performed on the results of the experiment after the first two cuttings in 1943, it would be for a standard split block design as shown above. However, what was done is that each alfalfa genotype plot or *wpeu* was divided into $b = 4$ split plot experimental units, *speus*. A randomly selected *speu* was then applied to one of $b = 4$ treatments as follows: B_0 was no third cutting, B_1 was a third cutting on September 1, B_2 was a third cutting on September 7, and B_3 was a third cutting on September 20. The yields presented are for first cutting yields in 1944. Without proper explanation of such a designed experiment, a person could think that periods over time are of a split plot or nested nature.

Example 2—Snedecor and Cochran (1980), Section 16.16, present data for an experiment on $a = 4$ dates of cutting asparagus, factor A. A randomized complete block experiment design with $r = 4$ replicates was used. The yields of asparagus over $b = 4$ years, factor B, are presented. Note that asparagus is a perennial crop. They analyze the data using trends over years as for some repeated measures experiments. However, since there is often an interest in the interaction of factor A with years, they could have analyzed the results of the experiment as follows.

Source of variation	Degrees of freedom
Total	$64 = abr$
Correction for mean	1
Replicates $= R$	$3 = r - 1$
Years $= B$	$3 = b - 1$
$R \times B =$ error B	$9 = (b-1)(r-1)$
Cutting dates $= A$	$3 = a - 1$
$R \times A =$ error A	$9 = (a-1)(r-1)$
$A \times B$	$9 = (a-1)(b-1)$
$R \times A \times B =$ error AB	$27 = (a-1)(b-1)(r-1)$

Example 3—Federer et al. (2002) describe a split block experiment design for $a = 2$ treatments of water, with and without zinc as factor A, and $b = 14$ time periods as factor B. There were $r = 3$ runs of the experiment. The response was number of fish out of five that migrated to the two levels of the factor A treatments. The time periods were crossed over the two treatments. This could also be considered as a repeated measurements experiment and trends studied if the interest is on trends rather than on interactions.

Example 4—Gomez and Gomez (1984) discuss a split block experiment design with $a = 6$ rice varieties, $b = 3$ levels of nitrogen application, and $r = 3$ replicates.

Example 5—Mead (1988) presents a split block design for an intercropping experiment with $a = 4$ cowpea varieties, $b = 5$ maize varieties, and $r = 3$ replicates.

Example 6—Large scale field experiments using an airplane to spread fertilizer and herbicides require that herbicides be crossed with fertilizers in a field. The fields are the replicates.

Example 7—In an educational setting, suppose that r groups of 30 students are to be used. Let $a = 5$ subjects (mathematics, english, social studies, language, and gymnastics) and $b = 6$ sex and social groupings. The order in which the students take the courses is in a random order for each of the r groups. The sex and social group is crossed with subjects. The response is a test score, for example. Examples of this nature often cause considerable confusion as to the nature of the experiment design and consequently about the statistical analysis. Note that all 30 students in a group stay the same across all five subjects.

Example 8—Hoshmand (1994) presents data for four levels of rainfall and four fertilizer treatments with five replicates. The rainfall levels are for two wet years and two drought years.

Experimenters and statisticians sometimes mistake a split block design for a split plot design. This is especially true when one of the factors involves time periods, say factor B, or other forms of repeated measurements through time, which they designate as split plot treatments. Whenever the times are *calendar or clock times*, the time periods are crossed over the other set of treatments, say A. This means that they are in a split block arrangement. Such experiments could be considered to be in a repeated measurement category but in many instances, each of the time periods and especially the interactions of factor A with time periods B are of interest to an experimenter, for example, the various pickings of a tomato or bean crop or the various cuttings of a hay crop.

There are instances when time periods can be split plot treatments. For example, in a forage crop experiment with alfalfa, say, the time to cut or harvest a crop is determined for *each experimental unit* by some characteristic like percent of plants in bloom. A *biological calendar, or clock*, is used for each experimental unit, eu, to determine when to harvest an eu. Not making a distinction between a calendar clock and a biological clock has made individuals commit the error of analyzing a split block design as a split plot design. Some of the concepts are illustrated in the examples listed below.

2.3. ANALYSIS OF VARIANCE

For the standard form of a split block design, the usual response model equation is

$$Y_{hij} = \mu + \rho_h + \alpha_i + \eta_{hi} + \beta_j + \delta_{hj} + \alpha\beta_{ij} + \varepsilon_{hij}, \qquad (2.1)$$

where $h = 1, 2, \ldots, r$, $i = 1, 2, \ldots, a$; and $j = 1, 2, \ldots, b$,

Y_{hij} is the response for the *hij*th experimental unit,

μ is a general mean effect,

ρ_h is the *h*th block or replicate effect and is identically and independently distributed with mean zero and variance σ_ρ^2, IID$(0, \sigma_\rho^2)$,

α_i is the effect of the *i*th level of factor A,

η_{hi} is a random error effect for factor A and η_{hi} is identically and independently distributed as IID$(0, \sigma_\eta^2)$,

β_j is the effect of the *j*th level of factor B,

δ_{hj} is a random error effect for factor B and δ_{hj} is identically and independently distributed IID$(0, \sigma_\delta^2)$,

$\alpha\beta_{ij}$ is the *ij*th interaction effect of the two factors A and B, and

ε_{hij} is a random error effect for the interaction effects and is identically and independently distributed as IID$(0, \sigma_\varepsilon^2)$.

The different random effects ρ_h, η_{hi}, δ_{hj}, and ε_{hij} are assumed to be independent.

The expected values of the various mean squares in the ANOVA table as described above in Section 2.1, for fixed A and B factors, and for the above response model equation (2.1) are as follows:

Source of variation	Degrees of freedom	Expected value of mean square
Total	rab	
Correction for mean	1	
Replicates = R	$r-1$	$\sigma_\varepsilon^2 + b\sigma_\eta^2 + a\sigma_\delta^2 + ab\sigma_\rho^2$
Factor B	$b-1$	$\sigma_{\alpha\varepsilon}^2 + a\sigma_\delta^2 + f(B)$
$R \times B$ = error B	$(r-1)(b-1)$	$\sigma_\varepsilon^2 + a\sigma_\delta^2$
Factor A	$a-1$	$\sigma_\varepsilon^2 + b\sigma_\eta^2 + f(A)$
$R \times A$ = error A	$(r-1)(a-1)$	$\sigma_\varepsilon^2 + b\sigma_\eta^2$
$A \times B$	$(a-1)(b-1)$	$\sigma_\varepsilon^2 + f(AB)$
$R \times A \times B$ = error AB	$(r-1)(a-1)(b-1)$	σ_ε^2

The functions of the A effects, the B effects, and the interaction AB effects are defined as $f(A) = rb \sum_{i=1}^{a} \alpha_i^2/(a-1)$, $f(B) = ra \sum_{j=1}^{b} \beta_j^2/(b-1)$, and $f(AB) = r \sum_{i=1}^{a} \sum_{j=1}^{b} \alpha\beta_{ij}^2/(a-1)(b-1)$, respectively.

2.4. F-TESTS

Adding the requirement of normality to the distributions for the random error effects described for equation (2.1), the following F-tests are appropriate for fixed effects for both factors. Let E_a equal the Error A mean square, E_b equal the Error B mean square, E_{ab} equal the interaction error mean square, MSA equal the factor A mean square, MSB equal the factor B mean square, and MSAB equal the interaction mean square. To test the hypothesis of no factor A effects, the F-test is the ratio of the Factor A mean square over the error A mean square, $F = MSA/E_a$. To test the hypothesis of no B effects, divide the Factor B mean square by the error B mean square to obtain the F-test, $F = MSB/E_b$. The F-test for no interaction, $F = MSAB/E_{ab}$, is obtained by dividing the interaction mean square by the error AB mean square. Each of the three F-tests requires a different error term. The experimental units are different for factor A, for factor B, and for the interaction of factors A and B.

The expected values of the mean squares in an analysis of variance for random factor A and replicate (block R) effects and for a fixed effect factor B would be as follows:

Source of variation	Degrees of freedom	Expected values of mean squares A and R random, B fixed
Block = R	$r-1$	$\sigma_\varepsilon^2 + a\sigma_\delta^2 + b\sigma_\eta^2 + ab\sigma_\rho^2$
Factor A	$a-1$	$\sigma_\varepsilon^2 + b\sigma_\eta^2 + rb\sigma_\alpha^2$
$A \times R$	$(a-1)(r-1)$	$\sigma_\varepsilon^2 + b\sigma_\eta^2$
B	$b-1$	$\sigma_\varepsilon^2 + a\sigma_\delta^2 + \frac{rb\sigma_{\alpha\beta}^2}{(b-1)} + f(\beta)$
$B \times R$	$(b-1)(r-1)$	$\sigma_\varepsilon^2 + a\sigma_\delta^2$
$A \times B$	$(a-1)(b-1)$	$\sigma_\varepsilon^2 + \frac{rb\sigma_{\alpha\beta}^2}{(b-1)}$
$A \times B \times R$	$(a-1)(b-1)(r-1)$	σ_ε^2

As can be seen from the above table of expected values of mean squares, there is no appropriate error term to test for factor B effects. Therefore, it is necessary to construct an error term and to determine the approximate number of degrees of freedom for the constructed error mean square. The remaining tests proceed as described above. The estimated error mean square for testing significance of the factor B effects is $E_b + \text{MSAB} - E_{ab}$. The degrees of freedom may be approximated by

$$\frac{(E_b + \text{MSAB} - E_{ab})^2}{E_b^2/(b-1)(r-1) + \text{MSAB}^2/(a-1)(b-1) + E_{ab}^2/(a-1)(b-1)(r-1)}. \quad (2.1)$$

The above is based on the Satterthwaithe method for approximating the degrees of freedom

2.5. STANDARD ERRORS FOR CONTRASTS OF EFFECTS

Denote the three error mean squares as E_a, E_b, and E_{ab} for factor A, for factor B, and for their interaction, respectively. The estimated standard error of a difference between two factor A means or effects, i and i', is estimated as

$$\text{SE}(\bar{y}_{i.} - \bar{y}_{i'.}) = \sqrt{\frac{2E_a}{rb}}. \quad (2.2)$$

The standard error of a difference between two factor B means or effects, j and j' is estimated as

$$\text{SE}(\bar{y}_{.j} - \bar{y}_{.j'}) = \sqrt{\frac{2E_b}{ra}}. \quad (2.3)$$

The standard error of a difference between two means at one level of factor A but different levels of factor B is estimated as (e.g., see, Mead, 1988)

$$\text{SE}(\bar{y}_{ij} - \bar{y}_{ij'}) = \sqrt{\frac{2\{E_b + (a-1)E_{ab}\}}{ar}}. \quad (2.4)$$

The standard error of a difference between means at two levels of factor A for the same level of factor B is estimated as

$$\text{SE}(\bar{y}_{ij} - \bar{y}_{i'j}) = \sqrt{\frac{2\{E_a + (b-1)E_{ab}\}}{br}}. \quad (2.5)$$

The standard error of a difference between two means for combinations ij and fg, where i is not equal to f or g and j is not equal to f or g is estimated by

$$\text{SE}(\bar{y}_{ij} - \bar{y}_{fg}) = \sqrt{\frac{2\{aE_a + bE_a + (ab - a - b)E_{ab}\}}{rab}}. \quad (2.6)$$

The variance of a mean for one level of factor A is

$$V(\bar{y}_{i.}) = \frac{\sigma_\varepsilon^2 + b\sigma_\eta^2 + b\sigma_\rho^2 + \sigma_\delta^2}{rb}. \tag{2.7}$$

The variance of a factor B level mean is

$$V(\bar{y}_{.j}) = \frac{\sigma_\varepsilon^2 + a\sigma_\delta^2 + a\sigma_\rho^2 + \sigma_\eta^2}{ra}. \tag{2.8}$$

The variance of a mean for combination ij is

$$V(\bar{y}_{.ij}) = \frac{\sigma_\varepsilon^2 + \sigma_\delta^2 + \sigma_\eta^2 + \sigma_\rho^2}{r}. \tag{2.9}$$

Nair (1944) presents a discussion of the above as well as some comments on tests of significance.

2.6. NUMERICAL EXAMPLES

Example 2.1—An experiment involving ten maize hybrids, numbered 0 to 9, were arranged in a randomized complete block experiment design with $a = 10$ hybrids and $r = 2$ replicates or complete blocks. Then, $b = 3$ generations, 1, 2, and 3, of the hybrids were laid out across the ten hybrids in each block. The experiment design for the generations was a randomized complete block with $b = 3$ generations and $r = 2$ replicates. The data for the response variable yield of ear corn and the field arrangement are given below (from Leonard and Clark, 1939).

					Block 1					
					Hybrid					
Generation	3	8	2	1	6	7	0	9	4	5
1	48	46	46	42	43	47	48	46	46	49
3	46	45	44	46	45	49	45	48	48	49
2	43	42	42	44	44	47	45	47	47	48

					Block 2					
					Hybrid					
Generation	4	3	9	5	1	7	2	8	6	0
2	46	45	46	45	43	48	44	44	47	43
3	48	44	46	45	50	51	48	46	48	43
1	42	42	44	43	44	48	47	46	44	42

NUMERICAL EXAMPLES 47

The SAS code and data for this experiment are given in Appendix 2.1. The various means are as follows:

	Hybrid										
Generation	0	1	2	3	4	5	6	7	8	9	Mean
1	45.0	43.0	46.5	45.0	44.0	46.0	43.5	47.5	46.0	45.0	45.2
2	44.0	43.5	43.0	44.0	46.5	46.5	45.5	47.5	43.0	46.5	45.0
3	44.0	48.0	46.0	45.0	48.0	47.0	46.5	50.0	45.5	47.0	46.7
Mean	44.3	44.8	45.2	44.7	46.2	46.5	45.2	48.3	44.8	46.2	45.6

An analysis of variance table and associated F-values for the above data set are as follows:

Source of variation	Degrees of freedom	Sum of squares	Mean square	F-value	prob > F
Total	60	125,151			
Correction for mean	1	124,852.82			
Block (replicate), R	1	2.82	2.82		
Hybrid, A	9	77.68	8.63	0.96	
$A \times R$, Error A	9	81.02	9.00		
Generation, B	2	35.43	17.72	2.18	
$B \times R$, Error B	2	16.23	8.12		
Generation × hybrid	18	61.57	3.42	2.63	0.02
$A \times B \times R$, Error AB	18	23.43	1.30		

Based on the above analysis of variance table, the generation by hybrid interaction is significant at the 2% level. Before drawing any conclusions, it would be wise to consult with the experimenter to ascertain if interaction was suspected. The interaction graph is displayed in Figure 2.1. Here we see that hybrids 1, 4, and 7 have the highest yields for generation 3 but this is not the case for the other generations. This explains a part of the apparent interaction.

Example 2.2—The example used here is not a standard split block design but is included to demonstrate some of the analytical considerations needed for an analysis of the data from such an experiment design. An experiment was conducted using five apple tree rootstocks, factor A, arranged in a 5 × 5 Latin square experiment design. Four soil treatments, factor B, were arranged in a randomized complete block experiment design using the five columns of the Latin square as the blocks. The rows and rootstocks of the Latin square are crossed with soil treatments. The data in Table 2.1 were obtained via simulation as the original data were not available at the time the statistical analysis was explained to the experimenter. The third factor added to this experiment will be explained in a later chapter dealing with extensions of the split block experiment design. These data are for one level of the third factor. A SAS code, data format, and output are presented in Appendix 2.2.

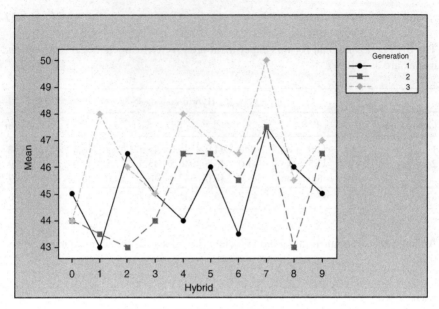

Figure 2.1. Hybrid by generation interaction.

Table 2.1. Simulated Data for Five Apple Tree Rootstocks Laid Out in a Latin Square Experiment Design with Four Soil Treatments Randomly Arranged in Each of the Five Columns of the Latin Square. Soil Treatments go Across All Rows.

Soil treatment	Column 1	Column 2	Column 3	Column 4	Column 5
1	R1 1027.85	R2 1004.33	R3 992.57	R4 994.60	R5 1019.89
2	982.74	977.86	993.71	1021.81	1017.48
3	1007.24	999.15	1012.57	995.03	987.82
4	1008.47	990.86	968.25	1002.17	995.63
1	R5 994.81	R3 1021.61	R4 1028.78	R1 996.18	R2 996.61
2	999.91	1014.46	1006.01	981.96	1011.94
3	1010.29	980.03	1015.04	985.63	972.76
4	1018.49	1014.80	1000.72	965.80	1011.99
1	R2 1013.52	R4 1010.08	R5 1004.83	R3 1003.92	R1 985.72
2	1017.40	997.66	983.86	999.33	1012.60
3	996.63	1012.12	1018.60	995.70	984.62
4	989.91	1019.53	1020.95	988.14	973.47
1	R3 990.17	R1 991.79	R2 1004.52	R5 1011.02	R4 1006.13
2	972.21	979.47	996.53	982.79	1005.57
3	1002.17	1004.70	1016.95	1018.23	1003.18
4	1017.56	1032.75	983.79	976.68	992.21
1	R4 985.12	R5 967.39	R1 993.54	R2 985.04	R3 1012.14
2	984.14	1009.78	1006.80	987.54	999.32
3	1010.74	1027.49	1001.24	990.53	1005.51
4	1004.63	1001.61	1010.73	982.68	998.86

Table 2.2. ANOVA Obtained from SAS PROC GLM output for Data of Table 2.1.

Source of variation	DF	Type I SS	Mean square
Total corrected for mean	99	22,310.37	
Column	4	1318.63	329.66
Rootstock	4	1159.76	289.94
Column × rootstock	16	2808.57	175.54
Soil treatment	3	351.88	117.29
Column × soil treatment	12	3863.28	321.94
Rootstock × soil treatment	12	825.96	68.83
Column × rootstock × soil treatment	48	11,982.28	249.63
Source of variation	DF	Type III SS	Mean square
Column	4	1318.63	329.66
Rootstock	4	1159.76	289.94
Column × rootstock	16	2808.57	175.54
Soil treatment	3	351.88	117.29
Column × soil treatment	12	3863.28	321.94
Rootstock × soil treatment	12	825.96	68.83
Column × rootstock × soil treatment	48	11,982.28	249.63

If the row blocking is ignored, the design is a standard split block experiment design with the columns being the blocks of a randomized complete block experiment design. The ANOVA for this arrangement is given in Table 2.2. Note that Type I and Type III sums of squares (SS) are identical as they should be for an orthogonal arrangement of factors. Also the degrees of freedom (DF) are as they should be for this design. A SAS PROC GLM code for the appropriate analysis including the row blocking is not available. Therefore, an additional run needs to be made to obtain the row sum of squares. Then, one may subtract this sum of squares from the column × rootstock interaction sum of squares. The result is the error sum of squares for rootstock. The column by rootstalk interaction sum of squares with 16 degrees of freedom is 2,808.57 (Table 2.2) and the row sum of squares with four degrees of freedom is 147.44 (Table 2.3). The error for rootstock sum of squares is 2808.57 − 147.44 = 2661.13 with 12 degrees of freedom and a mean square of 221.76.

The trouble is that rows run across the soil treatments. This has the effect of completely confounding the row and row × rootstock effects with the interactions of rootstock × soil treatment and column × rootstock × soil treatment as shown in Type III sum of squares in Table 2.3. The SAS/GLM procedure can give the wrong degrees of freedom as is indicated by interchanging the last two terms in the model of Table 2.4. The degrees of freedom for the column × rootstock Type I sum of squares should be 12 and not 16. Other models for this data set produced similar problems such as indicating 60 degrees of freedom for the three factor interaction column × rootstock × soil treatment instead of the correct number 48. The reason for this is unknown.

Table 2.3. ANOVA for a Second Model.

Source of variation	DF	Type I SS	Mean square
Total corrected for mean	99	22,310.37	
Row	4	147.44	36.86
Rootstock	4	1159.76	289.94
Row × rootstock	16	3979.76	248.74
Soil treatment	3	351.88	117.29
Column × soil treatment	12	3863.28	321.94
Rootstock × soil treatment	12	8275.96	68.83
Column × rootstock × soil treatment	48	11,982.28	249.63

Source of variation	DF	Type III SS	Mean square
Row	0	0.00	
Rootstock	4	1159.76	289.94
Row × rootstock	0	0.00	
Soil treatment	3	351.88	117.29
Column × soil treatment	12	3863.28	321.94
Rootstock × soil treatment	12	8275.96	68.83
Column × rootstock × soil treatment	48	11,982.28	249.63

Various F-tests may be made using the SAS PROC GLM and different error terms defined with commands placed after the model statement. These are given in the code presented in Appendix 2.2. For this data set, the F-statistics are near one as this is a simulated set of data and no significant effects are expected.

Table 2.4. ANOVA for a Third Model.

Source of variation	DF	Type I SS	Mean square
Total corrected for mean	99	22,310.37	
Row	4	147.44	36.86
Rootstock	4	1159.76	289.94
Soil treatment	3	351.88	117.29
Rootstock × soil treatment	12	825.96	68.83
Row × rootstock	16	3979.96	248.74
Column × soil treatment	12	3863.28	321.94
Error	48	11,982.28	249.63

Source of variation	DF	Type III SS	Mean square
Row	4	0.00	
Rootstock	4	1159.76	289.94
Soil treatment	3	351.88	117.29
Rootstock × soil treatment	12	825.96	68.83
Row × rootstock	12	0.00	0.00
Column × soil treatment	12	3863.28	321.94

The F-test for testing for significance of rootstock effects is

$$F = \frac{289.94}{221.76} = 1.31.$$

An F-test for testing for significance of soil treatment effects is

$$F = \frac{117.29}{321.94} = 0.36.$$

An F-test for testing significance of the rootstock by soil treatment interaction effects is

$$F = \frac{68.83}{249.63} = 0.28.$$

A standard error of a difference between two rootstock means, $\bar{y}_{..i.} - \bar{y}_{..i'.}$, is

$$\sqrt{\frac{2(221.76)}{4(5)}} = 4.71.$$

A standard error of a difference between two soil treatment means, $\bar{y}_{...j} - \bar{y}_{...j'}$, is

$$\sqrt{\frac{2(321.94)}{5(5)}} = 5.07.$$

A standard error of a difference between two soil treatment means for one rootstock, $\bar{y}_{..ij} - \bar{y}_{..ij'}$, {equation (2.4)} is

$$\sqrt{\frac{2\{321.94 + (5-1)(249.63)\}}{5(5)}} = 10.27.$$

A standard error of a difference between two means for one rootstock and one soil treatment and a mean from a different rootstock and a different soil treatment, $\bar{y}_{..ij} - \bar{y}_{..i'j'}$, {equation (2.6)} is

$$\sqrt{\frac{2\{5(221.76) + 4(321.94) + (5(4) - 5 - 4)(249.63)\}}{4(5)(5)}} = 10.14.$$

A standard error of a difference between two rootstock means at one level of soil treatment, $\bar{y}_{..ij} - \bar{y}_{..i'j}$ {equation (2.5)} is

$$\sqrt{\frac{2\{221.76 + (4-1)(249.63)\}}{4(5)}} = 9.85.$$

Table 2.5. Soil Treatment, Rootstock, and Soil Treatment by Rootstock Means for the Data of Table 2.1.

Soil treatment	Rootstock					
	1	2	3	4	5	mean
1	999.0	1000.8	1004.1	1004.9	999.6	1001.7
2	992.7	998.3	995.8	1003.0	998.8	997.7
3	996.7	995.2	999.2	1007.2	1012.5	1002.2
4	998.2	991.8	997.5	1003.9	1002.7	998.8
mean	996.7	996.5	999.2	1004.8	1003.4	1000.1

The soil treatment, the rootstock, and the rootstock by soil treatment means are presented in Table 2.5.

2.7. MULTIPLE COMPARISONS

Federer and McCulloch (1984) describe multiple comparisons for an experiment designed as a split block. All possible differences among the rootstock means are given in the top part of Table 2.6. A 95% least significant difference (lsd) is computed as

$$t_{0.05, df=12} \sqrt{\frac{2 \, \text{Error} \, R}{br}} = 2.179 \sqrt{\frac{2(221.76)}{4(5)}} = 2.179(4.71) = 10.26. \quad (2.10)$$

Error R equal Error A for rootstock means is

$$\frac{\text{column} \times \text{rootstock} - \text{row sums of squares}}{12} = \frac{2808.57 - 147.44}{12} = 221.76. \quad (2.11)$$

Table 2.6. Differences Among Rootstock Means and Among Soil Treatment Means.

Rootstock mean	4 1004.8	5 1003.4	3 999.2	1 996.7	2 996.5
2	8.3	6.9	2.7	0.2	
1	8.1	6.7	2.5		
3	5.6	4.2			
5	1.4				
Soil treatment mean	3 1002.2	1 1001.7	4 998.9	2 997.7	
2	4.5	4.0	1.1		
4	3.4	2.9			
1	0.5				

PRECISION   53

All differences are less than the value 10.26 in equation (2.10). A 95% honestly significant difference or Tukey's studentized multiple range test is computed as

$$q_{0.05, v=5, df=12}\sqrt{\frac{\text{Error } R}{br}} = 4.51(3.33) = 15.02. \tag{2.12}$$

Dunnett's test for comparing all rootstock means with a check, say number 1, at the 95% level is computed as

$$d_{0.05, df=12, a-1=4}\sqrt{\frac{2\text{ Error } R}{br}} = 2.88(4.71) = 13.56. \tag{2.13}$$

It is straightforward to compute these multiple comparison procedures as well as the esd and ssd (See Chapter 1) for soil treatment means and for the soil treatment by rootstock means, respectively, using the standard errors presented above in Section 2.5.

2.8. ONE REPLICATE OF A SPLIT BLOCK DESIGN

If an experimenter uses only one replicate of a split block experiment design, the following ANOVA table is appropriate:

Source of variation	Degrees of freedom
Total	ab
Correction for mean	1
Replicate $= R$	0
Factor A	$a - 1$
$R \times A =$ error A	0
Factor B	$b - 1$
$R \times B =$ error B	0
$A \times B$	$(a-1)(b-1)$
$R \times A \times B =$ error AB	0

As can be observed, there are no estimates of error A, error B, or error AB. When the situation is such, the experimenter often uses the $A \times B$ interaction as an error term for testing the significance of the factor A and factor B effects. For fixed effects A and B, such a procedure is not valid when the interaction of factor A and factor B is greater than zero as this would result in using an error mean square that is too large.

2.9. PRECISION

Following the results given in Chapter 1 for precision estimates for a split plot experiment design, similar formulae may be obtained for a split block experiment design. The estimated precision for rootstocks means in Example 2.2 is

$$\frac{(a-1)E_a + a(b-1)E_{ab}}{(ab-1)E_a} = \frac{(5-1)(221.76) + 5(4-1)(249.63)}{(5(4)-1)(221.76)} = 1.10. \quad (2.14)$$

The fact that the precision exceeds one is because the estimated error A is less than the estimated error AB. The estimated precision for the B or soil treatment means is

$$\frac{(b-1)E_b + b(a-1)E_{ab}}{(ab-1)E_b} = \frac{(4-1)(321.94) + 4(5-1)(249.63)}{(4(5)-1)(321.94)} = 0.81. \quad (2.15)$$

A measure of the precision for the interaction effects is

$$\frac{\frac{(a-1)E_a}{2} + \frac{(b-1)E_b}{2} + a(b-1)E_{ab}}{(ab-1)E_{ab}} =$$

$$\frac{\frac{(5-1)(221.76)}{2} + \frac{(4-1)(321.94)}{2} + 5(4-1)(249.63)}{(5(4) - 1 - 1/2)(249.63)} = 1.01. \quad (2.16)$$

Note that the sum of the weights for the means squares is $(5-1)/2 + (4-1)/2 + 5(4-1) = 18.5$ and that the denominator weight is $5(4) - 1 - 1/2 = 18.5$. These are equal as they should be. A weighted average of error A and error B mean squares is used here.

2.10. COMMENTS

Some analysts may desire to use a multiple comparisons procedure for comparing the factor A by factor B means or the factor A by factor B by factor C means for the experiment designs discussed in this and the previous chapter. Computer codes for doing this will be difficult, if not impossible, to write owing to the number of different standard errors of a difference of two means. For a standard split block designed experiment, five different standard errors of a difference between two means were given in equations (2.2)–(2.6). Even when using a multiple comparisons procedure for factor A and factor B means, it will be a programming problem to use the correct standard errors of a difference between two means {equations (2.2) and (2.3)}. A code statement, such as LSMEANS A*B/PDIFF, will make use of the single mean square listed as ERROR on the output. This is correct for some pairs but not for others.

In some cases, analysts may consider using PROC MIXED for the random effects in a model equation. When the effects are all orthogonal in an analysis of variance, i.e., Type III and Type III (or IV) sums of squares are identical, no additional information is obtained by using a fixed effects analysis. SAS GLM and SAS MIXED means are identical for orthogonal designs. Software using a mixed effects procedure should be used with caution as it is not always obvious to the analyst what the results mean or how they were obtained. It is not too difficult to find the output from a code that is incorrect for the purposes of the analysis of the data set under consideration, such as the incorrect number of degrees of freedom, incorrect F-test, and others.

2.11. PROBLEMS

Problem 2.1—For the data in Appendix 2.3 and for $P = 1$, obtain the analyses as described for Example 2.2 and discuss the results.

Problem 2.2—For the data in Appendix 2.3 and for $P = 2$, obtain the analyses as described for Example 2.2 and discuss the results.

Problem 2.3—Obtain the residuals for the analyses in Problems 2.1 and 2.2. Perform a Tukey stem and leaf plot to determine if there are outliers.

Problem 2.4—For the data given by Hoshmand (1994) on pages 174–175, obtain an analysis of variance for the data set. Write a code for obtaining an analysis of variance, means, and standard errors of means.

2.12 REFERENCES

Cochran, W. G. and G. M. Cox. (1950). *Experimental Designs*, John Wiley & Sons, Inc., New York, pp. 1–454.

Federer, W. T. (1955). *Experimental Design: Theory and Application*, The Macmillan Company, New York, pp. 1–544.

Federer, W. T. and C. E. McCulloch. (1984). Multiple comparisons procedures for some split plot and split block designs. In *Design of Experiments: Ranking and Selection (Essays in Honor of Robert E. Bechhoffer)* (Editors: T. J. Santner and A. C. Tamhane), Marcel Dekker, Inc., New York, pp. 7–22.

Federer, W. T., R. C. Lloyd, H. G. Ketola, and S. Qureshi. (2002). Selecting error mean squares in non-replicated and complex design situations. *BU-1595-M in the Technical Report Series of the Department of Biological Statistics and Computational Biology*, Cornell University, Ithaca, New York.

Gomez, K. A. and A. A. Gomez. (1984). *Statistical Procedures for Agricultural Research*, John Wiley & Sons, New York, Chichester, Brisbane, Toronto, pp. 1–680.

Hoshmand, A. R. (1994). *Experimental Research Design and Analysis: A Practical Approach for Agricultural and Natural Sciences*, CRC Press, Boca Raton, Ann Arbor, London, Tokyo, pp. 1–408.

Khargonkar, S. A. (1948). The estimation of missing plot value in split-plot and strip-plot trials. Journal of the Indian Society of Agricultural Statistics 1:147–161.

Lentner, M. and T. Bishop. (1986). *Experimental Design and Analysis*, Valley Book Company, Blacksburg, pp. 1–565.

Leonard, W. H. and A. G. Clark. (1939). *Field Plot Technique*, Burgess Publishing Company, Minneapolis, chapter 21.

Mead, R. (1988). *The Design of Experiments*, Cambridge University Press, Cambridge, New York, Port Chester, Melbourne, Sydney, pp. 1–620.

Nair, K. R. (1944). Calculation of standard errors and tests of significance of different types of treatment comparisons in split-plot and strip-plot arrangements of field experiments. Indian Journal of Agricultural Statistics 14:315–319.

Snedecor, G. W. and W. G. Cochran. (1980). *Statistical Methods*, seventh edition. The Iowa State University Press, Ames, Iowa, pp. 1–567.

Yates, F. (1933). The principles of orthogonality and confounding in replicated experiments. Journal of Agricultural Science 23:108–145.

APPENDIX 2.1. EXAMPLE 2.1 CODE

The SAS PROC GLM code and the data of Example 2.1 on the book's FTP site is given below:

```
data sbex;
input yield rep hyb gen;
datalines;
48 1 3 1
46 1 3 3
43 1 3 2
46 1 8 1
............
44 2 6 1
43 2 0 2
43 2 0 3
42 2 0 1
;
proc glm data = sbex;
   class rep hyb gen;
   model yield = rep hyb hyb*rep gen gen*rep gen*hyb;
   lsmeans hyb gen gen*hyb;
   Test H = hyb E = hyb*rep;
   Test H = gen E = gen*rep;
run;
```

APPENDIX 2.2. EXAMPLE 2.2 CODE

The SAS PROC GLM code and the data of Example 2.2 on the book's FTP site is given below:

```
data sbex2_2;
input row column R S Y;/*R = rootstock; S = soil treatment;
Y = response.*/
```

APPENDIX 57

```
datalines;
1 1 1 1 1027.85
1 1 1 2  982.74
1 1 1 3 1007.24
1 1 1 4 1008.47
1 2 2 1 1004.33

.............

5 2 5 2 1009.78
5 2 5 3 1027.49
5 2 5 4 1001.61
;
proc glm data = sbex2_2;
class row column R S;
model Y = column R R*column S S*column R*S/solution;
Test H = S E = S*column;
Test H = R*S E = Error;
lsmeans R S R*S/stderr;
run;

Proc glm data = sbex2_2;
class row column R S;
model Y = row R row*R S column*S R*S column*R*S;
run;

proc glm data = sbex2_2;
class row column R S;
model Y = row R S R*S row*R column*S column*R*S;
run;
```

An abbreviated form of the output from running the above code is given below.

The GLM procedure

Dependent variable: Y

Source	DF	Sum of squares	Mean square	F Value	Pr > F
Model	51	10,328.09037	202.51158	0.81	0.7688
Error	48	11,982.27509	249.63073		
Corrected total	99	22,310.36546			

R-Square	Coeff var	Root MSE	Y Mean
0.462928	1.579816	15.79971	1000.098

Source	DF	Type I SS	Mean square	F Value	Pr > F
Column	4	1318.629654	329.657414	1.32	0.2758
R	4	1159.764454	289.941113	1.16	0.3396
Column*R	16	2808.571226	175.535702	0.70	0.7766
S	3	351.882571	117.294190	0.47	0.7047

| Column*S | 12 | 3863.283714 | 321.940310 | 1.29 | 0.2555 |
| R*S | 12 | 825.958754 | 68.829896 | 0.28 | 0.9906 |

Source	DF	Type III SS	Mean square	F Value	Pr > F
Column	4	1318.629654	329.657414	1.32	0.2758
R	4	1159.764454	289.941113	1.16	0.3396
Column*R	16	2808.571226	175.535702	0.70	0.7766
S	3	351.882571	117.294190	0.47	0.7047
Column*S	12	3863.283714	321.940310	1.29	0.2555
R*S	12	825.958754	68.829896	0.28	0.9906

Tests of Hypotheses Using the Type III MS for column*S as an error term

Source	DF	Type III SS	Mean square	F Value	Pr > F
S	3	351.8825710	117.2941903	0.36	0.7800

Parameter		Estimate	Standard error	t Value	Pr > |t|
...					
R	1	−13.813000 B	14.13168726	−0.98	0.3332
R	2	−10.850500 B	14.13168726	−0.77	0.4464
...					
R*S	5 1	0.000000 B			
R*S	5 2	0.000000 B			
R*S	5 3	0.000000 B			
R*S	5 4	0.000000 B			

Least squares means

R		Y LSMEAN	Standard error	Pr > |t|
1		996.66500	3.53292	<0.0001
2		996.52700	3.53292	<0.0001
3		999.15150	3.53292	<0.0001
4		1004.76350	3.53292	<0.0001
5		1003.38250	3.53292	<0.0001
S		Y LSMEAN	Standard error	Pr > |t|
1		1001.69040	3.15994	<0.0001
2		997.71520	3.15994	<0.0001
3		1002.15880	3.15994	<0.0001
4		998.82720	3.15994	<0.0001
R	S	Y LSMEAN	Standard error	Pr > |t|
1	1	999.01600	7.06584	<0.0001
1	2	992.71400	7.06584	<0.0001
1	3	996.68600	7.06584	<0.0001
1	4	998.24400	7.06584	<0.0001
2	1	1000.80400	7.06584	<0.0001
2	2	998.25400	7.06584	<0.0001
2	3	995.20400	7.06584	<0.0001
2	4	991.84600	7.06584	<0.0001

APPENDIX

3	1	1004.08200	7.06584	<0.0001
3	2	995.80600	7.06584	<0.0001
3	3	999.19600	7.06584	<0.0001
3	4	997.52200	7.06584	<0.0001
4	1	1004.94200	7.06584	<0.0001
4	2	1003.03800	7.06584	<0.0001
4	3	1007.22200	7.06584	<0.0001
4	4	1003.85200	7.06584	<0.0001
5	1	999.60800	7.06584	<0.0001
5	2	998.76400	7.06584	<0.0001
5	3	1012.48600	7.06584	<0.0001
5	4	1002.67200	7.06584	<0.0001

Dependent variable: Y

Source	DF	Sum of squares	Mean square	F Value	Pr > F
Model	99	22,310.36546	225.35723		
Error	0	0.00000			
Corrected total	99	22,310.36546			

R-Square	Coeff Var	Root MSE	Y Mean
1.000000			1000.098

Source	DF	Type I SS	Mean square	F Value	Pr > F
Row	4	147.43859	36.85965		
R	4	1159.76445	289.94111		
Row*R	16	3979.76229	248.73514		
S	3	351.88257	117.29419		
Column*S	12	3863.28371	321.94031		
R*S	12	825.95875	68.82990		
Column*R*S	48	11,982.27509	249.63073		

Source	DF	Type III SS	Mean square	F Value	Pr > F
Row	0	0.00000			
R	4	1159.76445	289.94111		
Row*R	0	0.00000			
S	3	351.88257	117.29419		
Column*S	12	3863.28371	321.94031		
R*S	12	825.95875	68.82990		
Column*R*S	48	11,982.27509	249.63073		

Dependent variable: Y

Source	DF	Sum of squares	Mean square	F Value	Pr > F
Model	99	22,310.36546	225.35723		
Error	0	0.00000			
Corrected total	99	22,310.36546			

R-Square	Coeff var	Root MSE	Y Mean
1.000000			1000.098

Source	DF	Type I SS	Mean square	F Value	Pr > F
Row	4	147.43859	36.85965		
R	4	1159.76445	289.94111		
S	3	351.88257	117.29419		
R*S	12	825.95875	68.82990		
Row*R	16	3979.76229	248.73514		
Column*S	12	3863.28371	321.94031		
Column*R*S	48	11,982.27509	249.63073		

Source	DF	Type III SS	Mean square	F Value	Pr > F
Row	0	0.00000			
R	4	1159.76445	289.94111		
S	3	351.88257	117.29419		
R*S	12	825.95875	68.82990		
Row*R	0	0.00000			
Column*S	12	3863.28371	321.94031		
Column*R*S	48	11,982.27509	249.63073		

APPENDIX 2.3. PROBLEMS 2.1 AND 2.2 DATA

For Problems 2.1 and 2.2, the data set for a split block designed experiment is described below. The complete data set may be found on the book's FTP site. The combined analyses for this example will be described later in Chapter 4. The data are presented here to give two Latin square split block designed examples using $P = 1$ and $P = 2$ individually. P stands for previous crop, R denotes rootstock, and S denotes soil treatment. Height of plant is in centimeters.

```
data example;
input row P column R S height;
datalines;
1   1   1   3   4   103
1   1   1   3   2    98
1   1   1   3   3   101
1   1   1   3   1   101
1   1   2   4   2   100

......

5   2   4   3   4    83
5   2   5   2   1    99
5   2   5   2   2    96
5   2   5   2   3    98
5   2   5   2   4    99
;
```

CHAPTER 3

Variations of the Split Plot Experiment Design

3.1. INTRODUCTION

There are many variations of a split plot experiment design that arise as a result of the many goals and practices of an experimenter. The experimenter often appears not to think of the analysis that will be needed for a variation of the design, but only of the need or the desire for conducting an experiment in a certain manner. Difficulties may arise when comes the time to analyze the data. Some of the variations of the standard split plot experiment design that have been encountered during the course of consulting with experimenters are presented in the following sections and other variations in the following chapters.

When the split plot experimental unit is split into c split split plot experimental units (sspeus) for the c levels of a third factor C, a *split split plot experiment design* results. This design is discussed in the Section 3.2. The *split split split plot experiment design* results from splitting the sspeus into d split split split plot experimental units (ssspeus) for the d levels of a fourth factor, say factor D. This design is presented in Section 3.3. Situations arise in practice when the treatment design is not in a factorial arrangement. Two such situations are discussed in Section 3.4. As further variations, the split plot treatments may be arranged in an incomplete block experiment design within each whole plot treatment (Section 3.5) or in a row-column experiment design within each whole plot treatment or within each complete block (Section 3.6). In some situations the experimenter does not use randomization of treatments or feels it is not necessary, resulting in a systematic arrangement of the treatments in the experiment. The situation for a systematic arrangement of whole plot treatments is considered in Section 3.7 and for a systematic arrangement of split plot treatments in Section 3.8. In several instances, an experimenter has data from a

---

Variations on Split Plot and Split Block Experiment Designs, by Walter T. Federer and Freedom King
Copyright © 2007 John Wiley & Sons, Inc.

standard experiment design like a randomized complete block design. Then, for each experimental unit, the produce is split into the weights of various categories such as grass, legume, and weed or into categories such as number, grade, and quality. This results in a split plot design as discussed in Section 3.9. In Section 3.10, consideration of observational error versus experimental error is given for a particular split plot designed experiment. Measurements may be obtained for several time periods such as cuttings of a hay crop and pickings of a vegetable crop. The time dates may be selected in two ways. An illustration of a method for treating the two ways of determining time dates is discussed in Section 3.11. Some situations require an exploratory search for an appropriate model that describes the variation present in the experiment. Such a situation is discussed in Section 3.12 with a numerical example. The relationship between complete confounding in factorial experiments and a split plot design is discussed in Section 3.13. Additional comments are presented in Section 3.14.

3.2. SPLIT SPLIT PLOT EXPERIMENT DESIGN

A *split split plot experiment design* is a split plot design with the split plots divided into c split split plot experimental units to accommodate the c split split plot treatments of a third factor C. The treatment design is usually a three-factor factorial. Given that there are a whole plot treatments of factor A designed as a randomized complete block design in r replicates or complete blocks, b split plot treatments of factor B randomly allotted to the b split plot experimental units, speus, within each whole plot experimental unit, and the c split split plot treatments of factor C randomly allotted to the split split plot experimental units, sspeus, within each speu, there will be r randomizations for the factor A whole plot treatments, ra randomizations for the factor B split plot treatments, and rab randomizations for the factor C split split plot treatments. This means that there will be three different error terms when the three factors are considered to be fixed effects. The experimental units for the three factors are different and hence different error terms are required for each factor. There will be more error terms when one or more of the factors are considered to be random effects.

To describe the randomization procedure for a particular example of the experiment design described above, an example is used for $r = 3$ blocks, $a = 2$ levels of factor A, $b = 3$ levels of factor B, and $c = 6$ levels of factor C. The four steps in obtaining this plan are described below:

Step 1: Group similar experimental units into three blocks.

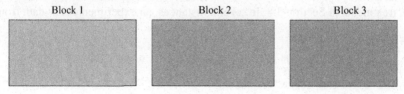

Step 2: Factor *A* levels randomly assigned to each block.

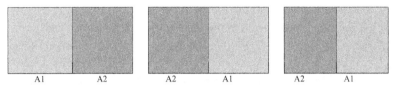

Step 3: Factor *B* levels randomly assigned to each level of Factor *A* and block.

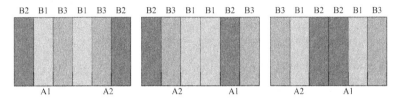

Step 4: Factor *C* levels randomly assigned to each level of Factor *B*, Factor *A*, and block.

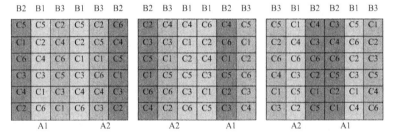

A linear model for the split split plot experiment design described above is

$$Y_{hijk} = \mu + \rho_h + \alpha_i + \delta_{hi} + \beta_j + \alpha\beta_{ij} + \lambda_{hij} + \gamma_k + \alpha\gamma_{ik} + \beta\gamma_{jk} + \alpha\beta\gamma_{ijk} + \varepsilon_{hijk} \quad (3.1)$$

$h = 1, \ldots, r, i = 1, \ldots, a, j = 1, \ldots, b,$ and $k = 1, \ldots, c,$

Y_{hijk} is the response (measurement) for the *hijk*th observation,

μ is a general mean effect,

ρ_h is an effect of the *h*th replicate randomly distributed with mean zero and variance σ_ρ^2,

α_i is the effect of the *i*th treatment (level) of factor *A*,

δ_{hi} is a random error effect distributed with mean zero and common variance σ_δ^2,

β_j is the effect of the *j*th treatment (level) of factor *B*,

$\alpha\beta_{ij}$ is the interaction effect of the *i*th level of *A* and the *j*th level of *B*,

λ_{hij} is a random error effect distributed with mean zero and common variance σ_λ^2,

γ_k is the effect of the *k*th level of factor *C*,

$\alpha\gamma_{ik}$ is an interaction effect for factors A and C,
$\beta\gamma_{jk}$ is an interaction effect for factors B and C,
$\alpha\beta\gamma_{ijk}$ is an interaction effect for the three factors A, B, and C, and
ε_{hijk} is a random error effect distributed with mean zero and common variance σ_ε^2.

The random effects ρ_h, δ_{hi}, λ_{hij}, and ε_{hijk} are assumed to be mutually independent.

A partitioning of the degrees of freedom in an analysis of variance table for a split-split-plot-designed experiment, as described above is as follows:

Source of variation	Degrees of freedom
Total	$rabc$
Correction for the mean	1
Replicate = R	$r-1$
Whole plot treatment = A	$a-1$
Error $A = R \times A$	$(r-1)(a-1)$
Split plot treatment = B	$b-1$
$A \times B$	$(a-1)(b-1)$
Error $B = R \times B$ within A	$a(r-1)(b-1)$
Split split plot treatment = C	$c-1$
$A \times C$	$(a-1)(c-1)$
$B \times C$	$(b-1)(c-1)$
$A \times B \times C$	$(a-1)(b-1)(c-1)$
Error $C = R \times C$ within A and B	$ab(r-1)(c-1)$

Considering only the fixed effects case for all three factors A, B, and C, the various standard errors of a difference between two means or effects is computed as follows, where $i \neq i'$, $j \neq j'$, $k \neq k'$, Error $A = E_a$, Error $B = E_b$, and Error $C = E_c$. The estimated standard error of a difference for comparing two means of factor A, $\bar{y}_{.i..} - \bar{y}_{.i'..}$, is

$$\mathrm{SE}(\bar{y}_{.i..} - \bar{y}_{.i'..}) = \sqrt{\frac{2E_a}{bcr}} \qquad (3.2)$$

for comparing two means of factor B, $\bar{y}_{..j.} - \bar{y}_{..j'.}$, is

$$\mathrm{SE}(\bar{y}_{..j.} - \bar{y}_{..j'.}) = \sqrt{\frac{2E_b}{acr}} \qquad (3.3)$$

for comparing two means of factor C, $\bar{y}_{...k} - \bar{y}_{...k'}$, is

$$\mathrm{SE}(\bar{y}_{...k} - \bar{y}_{...k'}) = \sqrt{\frac{2E_c}{abr}} \qquad (3.4)$$

for comparing two factor B means, $\bar{y}_{.ij.} - \bar{y}_{.ij'.}$, at the same level of factor A is

$$\text{SE}(\bar{y}_{.ij.} - \bar{y}_{.ij'.}) = \sqrt{\frac{2E_b}{cr}} \qquad (3.5)$$

for comparing two factor A means, $\bar{y}_{.ij.} - \bar{y}_{.i'j.}$, at the same level of factor B is

$$\text{SE}(\bar{y}_{.ij.} - \bar{y}_{.i'j.}) = \sqrt{\frac{2[E_b(b-1) + E_a]}{bcr}} \qquad (3.6)$$

for comparing two factor C means at the same level of factor A, $\bar{y}_{.i.k} - \bar{y}_{.i.k'}$, is

$$\text{SE}(\bar{y}_{.i.k} - \bar{y}_{.i.k'}) = \sqrt{\frac{2E_c}{rb}} \qquad (3.7)$$

for comparing two factor A means at the same level of factor C, $\bar{y}_{.i.k} - \bar{y}_{.i'.k}$, is

$$\text{SE}(\bar{y}_{.i.k} - \bar{y}_{.i'.k}) = \sqrt{\frac{2[E_c(c-1) + E_a]}{bcr}} \qquad (3.8)$$

for comparing two factor C means at the same level of factor B, $\bar{y}_{..jk} - \bar{y}_{..jk'}$, is

$$\text{SE}(\bar{y}_{..jk} - \bar{y}_{..jk'}) = \sqrt{\frac{2E_c}{ar}} \qquad (3.9)$$

for comparing two factor B means at the same level of factor C, $\bar{y}_{..jk} - \bar{y}_{..j'k}$, is

$$\text{SE}(\bar{y}_{..jk} - \bar{y}_{..j'k}) = \sqrt{\frac{2[E_c(c-1) + E_b]}{acr}} \qquad (3.10)$$

and for comparing two levels of factor A for the same level of factor B and for the same level of factor C, $\bar{y}_{.ijk} - \bar{y}_{.i'jk}$, is

$$\text{SE}(\bar{y}_{.ijk} - \bar{y}_{.i'jk}) = \sqrt{\frac{2[E_a(b-1) + E_b + b(c-1)E_a]}{rbc}} \qquad (3.11)$$

The number of degrees of freedom for the standard errors of a difference between two means or effects in equations (3.6), (3.8), (3.10), and (3.11) needs to be approximated as none of the mean squares in an analysis of variance are of the correct form for these standard errors.

Variances of a difference between two means are given below where the various means are $\bar{y}_{ij\cdot}, \bar{y}_{\cdot i\cdot k}$, and $\bar{y}_{\cdot\cdot jk}$, for $i \neq i', j \neq j'$, and $k \neq k'$.

$$\operatorname{Var}(\bar{y}_{ij\cdot} - \bar{y}_{i'j'\cdot}) = \frac{2(c\sigma_\delta^2 + c\sigma_\lambda^2 + \sigma_\varepsilon^2)}{rc} \quad (3.12)$$

$$\operatorname{Var}(\bar{y}_{ij\cdot} - \bar{y}_{i'j\cdot}) = \frac{2(c\sigma_\delta^2 + c\sigma_\lambda^2 + \sigma_\varepsilon^2)}{rc} \quad (3.13)$$

$$\operatorname{Var}(\bar{y}_{ij\cdot} - \bar{y}_{ij'\cdot}) = \frac{2(c\sigma_\lambda^2 + \sigma_\varepsilon^2)}{rc} \quad (3.14)$$

$$\operatorname{Var}(\bar{y}_{\cdot i\cdot k} - \bar{y}_{\cdot i'\cdot k'}) = \frac{2(b\sigma_\delta^2 + \sigma_\lambda^2 + \sigma_\varepsilon^2)}{rb} \quad (3.15)$$

$$\operatorname{Var}(\bar{y}_{\cdot i\cdot k} - \bar{y}_{\cdot i'\cdot k}) = \frac{2(b\sigma_\delta^2 + \sigma_\lambda^2 + \sigma_\varepsilon^2)}{rb} \quad (3.16)$$

$$\operatorname{Var}(\bar{y}_{\cdot i\cdot k} - \bar{y}_{\cdot i\cdot k'}) = \frac{2(\sigma_\varepsilon^2)}{rb} \quad (3.17)$$

$$\operatorname{Var}(\bar{y}_{\cdot\cdot jk} - \bar{y}_{\cdot\cdot j'k'}) = \frac{2(\sigma_\lambda^2 + \sigma_\varepsilon^2)}{ra} \quad (3.18)$$

$$\operatorname{Var}(\bar{y}_{\cdot\cdot jk} - \bar{y}_{\cdot\cdot j'k}) = \frac{2(\sigma_\lambda^2 + \sigma_\varepsilon^2)}{ra} \quad (3.19)$$

$$\operatorname{Var}(\bar{y}_{\cdot\cdot jk} - \bar{y}_{\cdot\cdot jk'}) = \frac{2(\sigma_\varepsilon^2)}{ra} \quad (3.20)$$

The expected values of the mean squares for the analysis of variance for factors A, B, and C, as fixed effects, and as random effects, are presented below:

Source	Fixed effect	Random effect
R	$\sigma_\varepsilon^2 + c\sigma_\lambda^2 + bc\sigma_\delta^2 + abc\sigma_\rho^2$	$\sigma_\varepsilon^2 + c\sigma_\lambda^2 + bc\sigma_\delta^2 + abc\sigma_\rho^2$
A	$\sigma_\varepsilon^2 + c\sigma_\lambda^2 + bc\sigma_\delta^2 + f(\alpha)$	$\sigma_\varepsilon^2 + c\sigma_\lambda^2 + bc\sigma_\delta^2 + r\sigma_{\alpha\beta\gamma}^2 + rb\sigma_{\alpha\gamma}^2 + rc\sigma_{\alpha\beta}^2 + rbc\sigma_\alpha^2$
Error A	$\sigma_\varepsilon^2 + c\sigma_\lambda^2 + bc\sigma_\delta^2$	$\sigma_\varepsilon^2 + c\sigma_\lambda^2 + bc\sigma_\delta^2$
B	$\sigma_\varepsilon^2 + c\sigma_\lambda^2 + f(\beta)$	$\sigma_\varepsilon^2 + c\sigma_\lambda^2 + r\sigma_{\alpha\beta\gamma}^2 + ar\sigma_{\beta\gamma}^2 + rc\sigma_{\alpha\beta}^2 + rac\sigma_\beta^2$
A × B	$\sigma_\varepsilon^2 + c\sigma_\lambda^2 + f(\alpha\beta)$	$\sigma_\varepsilon^2 + c\sigma_\lambda^2 + r\sigma_{\alpha\beta\gamma}^2 + rc\sigma_{\alpha\beta}^2$
Error B	$\sigma_\varepsilon^2 + c\sigma_\lambda^2$	$\sigma_\varepsilon^2 + c\sigma_\lambda^2$
C	$\sigma_\varepsilon^2 + f(\gamma)$	$\sigma_\varepsilon^2 + r\sigma_{\alpha\beta\gamma}^2 + ar\sigma_{\beta\gamma}^2 + rb\sigma_{\alpha\gamma}^2 + rab\sigma_\gamma^2$
A × C	$\sigma_\varepsilon^2 + f(\alpha\gamma)$	$\sigma_\varepsilon^2 + r\sigma_{\alpha\beta\gamma}^2 + rb\sigma_{\alpha\gamma}^2$
B × C	$\sigma_\varepsilon^2 + f(\beta\gamma)$	$\sigma_\varepsilon^2 + r\sigma_{\alpha\beta\gamma}^2 + ar\sigma_{\beta\gamma}^2$
A × B × C	$\sigma_\varepsilon^2 + f(\alpha\beta\gamma)$	$\sigma_\varepsilon^2 + r\sigma_{\alpha\beta\gamma}^2$
Error C	σ_ε^2	σ_ε^2

For fixed effect factors A, B, and C, there are three different error terms. Given the usual assumptions about F-tests in an analysis of variance, Error A, E_a, is used to test for factor A effects. Error B, E_b, is used to test for factor B effects and for the $A \times B$

interaction effects. The significance of factor C effects, the $A \times C$, the $B \times C$, and the $A \times B \times C$ effects are tested using the Error C mean square, E_c.

If the factor A effects are random and B and C are fixed effects, then the $A \times B$ interaction mean square is used in the F-test for factor B effects. The variance component $\sigma^2_{\beta\gamma}$ does not appear in the expected value for the factor B mean square as the summation over levels of factor C, is zero, that is $\sum_k \beta\gamma_{jk} = 0$. The $A \times C$ interaction mean square is used in the F-test for factor C effects. The $A \times B \times C$ interaction mean square is used in the F-test for testing for $A \times C$ and $B \times C$ interaction effects. The Error B mean square is used to test for the $A \times B$ interaction effects, as the variance component for the three factor interaction effects does not appear in the expected value.

Other situations can arise for a mixed model situation. For example, factor B could be a random effect and factors A and C fixed effects, factor C could be random and other effects fixed, factors A and B could be random and C fixed, factors A and C could be random and B fixed, or factors B and C could be random and A fixed. Appropriate adjustments in testing will need to be made for these situations.

3.3. SPLIT SPLIT SPLIT PLOT EXPERIMENT DESIGN

This design involves splitting (dividing) the split split plot experimental units into d split split split plot experimental units, ssspeus. The resulting design is a *split split split plot experiment design*. The d treatments of factor D are randomly allotted to the split split split plot experimental units within each of the treatment combinations for factors A, B, and C. Thus, there are $rabc$ randomizations of the factor D treatments. The treatment design considered here is a four-factor factorial treatment design. A linear model for this design is as follows:

$$Y_{ghijk} = \mu + \rho_g + \alpha_h + \pi_{gh} + \beta_i + \alpha\beta_{hi} + \lambda_{ghi} + \gamma_j + \alpha\gamma_{hj} + \beta\gamma_{ij} + \alpha\beta\gamma_{hij} + \eta_{ghij} + \delta_k$$
$$+ \alpha\delta_{hk} + \beta\delta_{ik} + \alpha\beta\delta_{hik} + \gamma\delta_{jk} + \alpha\gamma\delta_{hjk} + \beta\gamma\delta_{ijk} + \alpha\beta\gamma\delta_{hijk} + \varepsilon_{ghijk} \quad (3.21)$$

$g = 1, \ldots, r, h = 1, \ldots, a, i = 1, \ldots, b, j = 1, \ldots, c$, and $k = 1, \ldots, d$,

Y_{ghijk} is the response (measurement) for observation ghijk,

μ is a general mean effect,

ρ_g is the gth block effect randomly distributed with mean zero and variance σ^2_ρ,

α_h is the effect of the hth level of factor A,

π_{gh} is a random error effect distributed with mean zero and variance σ^2_π,

β_i is the effect of the ith level of factor B,

$\alpha\beta_{hi}$ is an interaction effect of level h for factor A with level i for factor B,

λ_{ghi} is a random error effect distributed with mean zero and variance σ^2_λ,

γ_j is the effect of level j for factor C,

$\alpha\gamma_{hj}$ is an interaction effect of level h for factor A and level j for factor C,

$\beta\lambda_{ij}$ is an interaction effect of level i for factor B and level j for factor C,

$\alpha\beta\lambda_{hij}$ is an interaction effect among level h of factor A, level i of factor B, and level j of factor C,

η_{ghij} is a random error effect distributed with mean zero and variance σ_η^2,

δ_k is the effect of level k for factor D,

$\alpha\delta_{hk}$ is an interaction effect of level h for factor A and level k for factor D,

$\beta\delta_{ik}$ is an interaction effect of level i for factor B and level k for factor D,

$\alpha\beta\delta_{hik}$ is an interaction effect among level h for factor A, level i for factor B, and level k for factor D,

$\gamma\delta_{jk}$ is an interaction effect of level j for factor C and level k for factor D,

$\alpha\gamma\delta_{hjk}$ is an interaction effect among level h of factor A, level j of factor C, and level k of factor D,

$\beta\gamma\delta_{ijk}$ is an interaction effect among level i for factor B, level j for factor C, and level k for factor D,

$\alpha\beta\gamma\delta_{hijk}$ is an interaction effect among level h of factor A, level i of factor B, level j of factor C, and level k for factor D,

ε_{ghijk} is a random error effect distributed with mean zero and variance σ_ε^2.

The random effects in the above model are assumed to be mutually independent.

A partitioning of the degrees of freedom in an analysis of variance table for this design follows:

Source of variation	Degrees of freedom
Total	$rabcd$
Correction for mean	1
Replicate $= R$	$r - 1$
Whole plot treatment $=$ Factor A	$a - 1$
$R \times A =$ Error A	$(a - 1)(r - 1)$
Split plot treatment $=$ Factor B	$b - 1$
$A \times B$	$(a - 1)(b - 1)$
$R \times B$ within $A =$ Error B	$a(r - 1)(b - 1)$
Split split plot treatment $=$ Factor C	$c - 1$
$A \times C$	$(a - 1)(c - 1)$
$B \times C$	$(b - 1)(c - 1)$
$A \times B \times C$	$(a - 1)(b - 1)(c - 1)$
$R \times C$ within A and $B =$ Error C	$ab(r - 1)(c - 1)$
Split split split plot treatment $=$ Factor D	$d - 1$
$A \times D$	$(a - 1)(d - 1)$
$B \times D$	$(b - 1)(d - 1)$
$C \times D$	$(c - 1)(d - 1)$
$A \times B \times D$	$(a - 1)(b - 1)(d - 1)$
$A \times C \times D$	$(a - 1)(c - 1)(d - 1)$
$B \times C \times D$	$(b - 1)(c - 1)(d - 1)$
$A \times B \times C \times D$	$(a - 1)(b - 1)(c - 1)(d - 1)$
$R \times D$ within $A, B,$ and $C =$ Error D	$abc(r - 1)(d - 1)$

SPLIT SPLIT SPLIT PLOT EXPERIMENT DESIGN

There are r randomizations for factor A, ra randomizations for factor B, rab randomizations for factor C, and $rabc$ randomizations for factor D. Thus, there are four different error terms Error A, Error B, Error C, and Error D. This leads to different standard errors for the many kinds of contrasts of means. Various estimates of standard errors of a difference between means for the fixed effects case are given below. The estimated standard error of a difference between two factor A means, $\bar{y}_{.h...} - \bar{y}_{.h'...}$, is given below; $h \neq h'$, $i \neq i'$, $j \neq j'$, and $k \neq k'$,

$$\text{SE}(\bar{y}_{.h...} - \bar{y}_{.h'...}) = \sqrt{\frac{2\text{Error A}}{rbcd}} \tag{3.22}$$

The standard error of a difference between two factor B means, $\bar{y}_{..i..} - \bar{y}_{..i'..}$, is

$$\text{SE}(\bar{y}_{..i..} - \bar{y}_{..i'..}) = \sqrt{\frac{2\text{Error B}}{racd}} \tag{3.23}$$

The standard error of a difference between two factor C means, $\bar{y}_{...j.} - \bar{y}_{...j'.}$, is

$$\text{SE}(\bar{y}_{...j.} - \bar{y}_{...j'.}) = \sqrt{\frac{2\text{Error C}}{rabd}} \tag{3.24}$$

The standard error of a difference between two factor D means, $\bar{y}_{....k} - \bar{y}_{....k'}$, is

$$\text{SE}(\bar{y}_{....k} - \bar{y}_{....k'}) = \sqrt{\frac{2\text{Error D}}{rabc}} \tag{3.25}$$

The standard error of a difference between two factor D means, $\bar{y}_{.hijk} - \bar{y}_{.hijk'}$, at the same level of factors A, B, and C, is

$$\text{SE}(\bar{y}_{.hijk} - \bar{y}_{.hijk'}) = \sqrt{\frac{2\text{Error D}}{r}} \tag{3.26}$$

The standard error of a difference between two factor D means, $\bar{y}_{.hi.k} - \bar{y}_{.hi.k'}$, at the same level of factors A and B is

$$\text{SE}(\bar{y}_{.hi.k} - \bar{y}_{.hi.k'}) = \sqrt{\frac{2\text{Error D}}{rc}} \tag{3.27}$$

The standard error of a difference between factor D means, $\bar{y}_{.h.jk} - \bar{y}_{.h.jk'}$, at the same level of factors A and C is

$$\text{SE}(\bar{y}_{.h.jk} - \bar{y}_{.h.jk'}) = \sqrt{\frac{2\text{Error D}}{rb}} \tag{3.28}$$

The standard error of a difference between two factor D means, $\bar{y}_{..ijk} - \bar{y}_{..ijk'}$, at the same level of factors B and C is

$$\text{SE}(\bar{y}_{..ijk} - \bar{y}_{..ijk'}) = \sqrt{\frac{2\text{Error D}}{ra}} \qquad (3.29)$$

The standard error of a difference between two factor D means, $\bar{y}_{.h..k} - \bar{y}_{.h..k'}$, at the same level of factor A is

$$\text{SE}(\bar{y}_{.h..k} - \bar{y}_{.h..k'}) = \sqrt{\frac{2\text{Error D}}{rbc}} \qquad (3.30)$$

The standard error of a difference between two factor D means, $\bar{y}_{...jk} - \bar{y}_{...jk'}$, at the same level of factor C is

$$\text{SE}(\bar{y}_{...jk} - \bar{y}_{...jk'}) = \sqrt{\frac{2\text{Error D}}{rab}} \qquad (3.31)$$

The standard error of a difference between two factor D means, $\bar{y}_{..i.k} - \bar{y}_{..i.k'}$, at the same level of factor B is

$$\text{SE}(\bar{y}_{..i.k} - \bar{y}_{..i.k'}) = \sqrt{\frac{2\text{Error D}}{rac}} \qquad (3.32)$$

The standard error of a difference between two factor C means, $\bar{y}_{.hij.} - \bar{y}_{.hij'.}$, at the same level of factors A and B is

$$\text{SE}(\bar{y}_{.hij.} - \bar{y}_{.hij'.}) = \sqrt{\frac{2\text{Error C}}{rd}} \qquad (3.33)$$

The standard error of a difference between two factor C means, $\bar{y}_{.h.j.} - \bar{y}_{.h.j'.}$, at the same level of factor A is

$$\text{SE}(\bar{y}_{.h.j.} - \bar{y}_{.h.j'.}) = \sqrt{\frac{2\text{Error C}}{rbd}} \qquad (3.34)$$

The standard error of a difference between two factor C means, $\bar{y}_{..ij.} - \bar{y}_{..ij'.}$, at the same level of factor B is

$$\text{SE}(\bar{y}_{..ij.} - \bar{y}_{..ij'.}) = \sqrt{\frac{2\text{Error C}}{rad}} \qquad (3.35)$$

SPLIT SPLIT SPLIT PLOT EXPERIMENT DESIGN

The standard error of a difference between two factor B means, $\bar{y}_{.hi..} - \bar{y}_{.hi'..}$, at the same level of factor A is

$$\text{SE}(\bar{y}_{.hi..} - \bar{y}_{.hi'..}) = \sqrt{\frac{2\text{Error B}}{rcd}} \tag{3.36}$$

Some additional variances of the difference between two means are given below. For comparing levels hi of factors A and B, the variances are as follows:

$$\text{Var}(\bar{y}_{.hi..} - \bar{y}_{.h'i'..}) = \frac{2(cd\sigma_\pi^2 + cd\sigma_\lambda^2 + d\sigma_\eta^2 + \sigma_\varepsilon^2)}{rcd} \tag{3.37}$$

$$\text{Var}(\bar{y}_{.hi..} - \bar{y}_{.hi'..}) = \frac{2(cd\sigma_\lambda^2 + d\sigma_\eta^2 + \sigma_\varepsilon^2)}{rcd} \tag{3.38}$$

$$\text{Var}(\bar{y}_{.hi..} - \bar{y}_{.h'i..}) = \frac{2(cd\sigma_\pi^2 + cd\sigma_\lambda^2 + d\sigma_\eta^2 + \sigma_\varepsilon^2)}{rcd} \tag{3.39}$$

To compare levels hj of factors A and C, the following variances pertain:

$$\text{Var}(\bar{y}_{.h.j.} - \bar{y}_{.h'.j'.}) = \frac{2(bd\sigma_\pi^2 + d\sigma_\lambda^2 + d\sigma_\eta^2 + \sigma_\varepsilon^2)}{rbd} \tag{3.40}$$

$$\text{Var}(\bar{y}_{.h.j.} - \bar{y}_{.h.j'.}) = \frac{2(d\sigma_\lambda^2 + d\sigma_\eta^2 + \sigma_\varepsilon^2)}{rbd} \tag{3.41}$$

$$\text{Var}(\bar{y}_{.h.j.} - \bar{y}_{.h'.j.}) = \frac{2(bd\sigma_\pi^2 + d\sigma_\lambda^2 + d\sigma_\eta^2 + \sigma_\varepsilon^2)}{rbd} \tag{3.42}$$

For comparing levels hk of factors A and D, use the following variances:

$$\text{Var}(\bar{y}_{.h..k} - \bar{y}_{.h'..k'}) = \frac{2(bc\sigma_\pi^2 + c\sigma_\lambda^2 + \sigma_\eta^2 + \sigma_\varepsilon^2)}{rbc} \tag{3.43}$$

$$\text{Var}(\bar{y}_{.h..k} - \bar{y}_{.h..k'}) = \frac{2(c\sigma_\lambda^2 + \sigma_\eta^2 + \sigma_\varepsilon^2)}{rbc} \tag{3.44}$$

$$\text{Var}(\bar{y}_{.h..k} - \bar{y}_{.h'..k}) = \frac{2(bc\sigma_\pi^2 + c\sigma_\lambda^2 + \sigma_\eta^2 + \sigma_\varepsilon^2)}{rbc} \tag{3.45}$$

To compare levels ij of factors B and C, use the following variances:

$$\text{Var}(\bar{y}_{..ij.} - \bar{y}_{..i'j'.}) = \frac{2(d\sigma_\lambda^2 + d\sigma_\eta^2 + \sigma_\varepsilon^2)}{rad} \tag{3.46}$$

$$\text{Var}(\bar{y}_{..ij.} - \bar{y}_{..ij'.}) = \frac{2(d\sigma_\eta^2 + \sigma_\varepsilon^2)}{rad} \tag{3.47}$$

$$\text{Var}(\bar{y}_{..ij.} - \bar{y}_{..i'j.}) = \frac{2(d\sigma_\lambda^2 + d\sigma_\eta^2 + \sigma_\varepsilon^2)}{rad} \tag{3.48}$$

The following variances are for comparing *ik* levels of factors *B* and *D*:

$$\text{Var}(\bar{y}_{..ik} - \bar{y}_{.i'k'}) = \frac{2(c\sigma_\lambda^2 + \sigma_\eta^2 + \sigma_\varepsilon^2)}{rac} \qquad (3.49)$$

$$\text{Var}(\bar{y}_{..ik} - \bar{y}_{..ik'}) = \frac{2\sigma_\varepsilon^2}{rac} \qquad (3.50)$$

$$\text{Var}(\bar{y}_{..ik} - \bar{y}_{..i'k}) = \frac{2(c\sigma_\lambda^2 + \sigma_\eta^2 + \sigma_\varepsilon^2)}{rac} \qquad (3.51)$$

For comparing levels *jk* of factors *C* and *D*, use the following variances:

$$\text{Var}(\bar{y}_{...jk} - \bar{y}_{...j'k'}) = \frac{2(\sigma_\eta^2 + \sigma_\varepsilon^2)}{rab} \qquad (3.52)$$

$$\text{Var}(\bar{y}_{...jk} - \bar{y}_{...jk'}) = \frac{2\sigma_\varepsilon^2}{rab} \qquad (3.53)$$

$$\text{Var}(\bar{y}_{...jk} - \bar{y}_{...j'k}) = \frac{2(\sigma_\eta^2 + \sigma_\varepsilon^2)}{rab} \qquad (3.54)$$

To compare levels *hij* of factors *A, B,* and *C*, use the following variances:

$$\text{Var}(\bar{y}_{.hij.} - \bar{y}_{.h'i'j'.}) = \frac{2(d\sigma_\pi^2 + d\sigma_\lambda^2 + d\sigma_\eta^2 + \sigma_\varepsilon^2)}{rd} \qquad (3.55)$$

$$\text{Var}(\bar{y}_{.hij.} - \bar{y}_{.hij'.}) = \frac{2(d\sigma_\eta^2 + \sigma_\varepsilon^2)}{rd} \qquad (3.56)$$

$$\text{Var}(\bar{y}_{.hij.} - \bar{y}_{.hi'j'.}) = \frac{2(d\sigma_\lambda^2 + d\sigma_\eta^2 + \sigma_\varepsilon^2)}{rd} \qquad (3.57)$$

$$\text{Var}(\bar{y}_{.hij.} - \bar{y}_{.hi'j.}) = \frac{2(d\sigma_\lambda^2 + d\sigma_\eta^2 + \sigma_\varepsilon^2)}{rd} \qquad (3.58)$$

$$\text{Var}(\bar{y}_{.hij.} - \bar{y}_{.h'i'j.}) = \frac{2(d\sigma_\pi^2 + d\sigma_\lambda^2 + d\sigma_\eta^2 + \sigma_\varepsilon^2)}{rd} \qquad (3.59)$$

$$\text{Var}(\bar{y}_{.hij.} - \bar{y}_{.h'ij'.}) = \frac{2(d\sigma_\pi^2 + d\sigma_\lambda^2 + d\sigma_\eta^2 + \sigma_\varepsilon^2)}{rd} \qquad (3.60)$$

$$\text{Var}(\bar{y}_{.hij.} - \bar{y}_{.h'ij.}) = \frac{2(d\sigma_\pi^2 + d\sigma_\lambda^2 + d\sigma_\eta^2 + \sigma_\varepsilon^2)}{rd} \qquad (3.61)$$

Variances similar to equations (3.55) to (3.61) may be developed for comparisons of levels *hik* of factors *A, B,* and *D*; for comparing levels *hjk* of factors *A, C,* and D; for comparing levels ijk of factors B, C, and D; and for comparing levels $hijk$ of factors A, B, C, and D. Most of these variances will need to be constructed using the solutions from the variance components. The solutions for variance components may be obtained from an analysis of variance, ANOVA, or from a mixed model procedure such as may be found in SAS PROC MIXED. The latter procedure assumes normality assumptions whereas the ANOVA procedure does not.

3.4. WHOLE PLOTS NOT IN A FACTORIAL ARRANGEMENT

Federer (1955) discusses an example when the treatments are not in a factorial arrangement. The field design is a split split plot experiment design with three replicates. Factor A has four rates of planting, factor B has three levels of fertilizer with one level being no fertilizer, and factor C has two methods of application. Obviously, there are not two levels of application for no fertilizer. In computing the $B \times C$ interaction the zero level needs to be omitted. If not omitted, an interaction could be indicated when in fact none exists. Instead of two degrees of freedom for this interaction, there is only one, the other degree of freedom involves the comparison of two experimental units with the same treatment and is relegated to the Error C mean square, as are three of the six degrees of freedom from the $A \times B \times C$, interaction, as shown in the analysis of variance table partitioning of the degrees of freedom as follows:

Source of variation	Degrees of freedom
Total	$72 = abcr$
Correction for the mean	1
Replicate $= R$	$2 = r - 1$
Rate of planting $= A$	$3 = a - 1$
Error $A = R \times A$	$6 = (a - 1)(r - 1)$
Levels of fertilizer $= B$	$2 = b - 1$
$A \times B$	$6 = (a - 1)(b - 1)$
Error B	$16 = a(b - 1)(r - 1)$
Methods of application $= C$	$1 = c - 1$
$A \times C$	$(3 = (a - 1)(c - 1)$
$B \times C$	$1 = (b - 2)(c - 1)$
$A \times B \times C$	$3 = (a - 1)(b - 2)(c - 1)$
Error C	$1 + 3 + 24 = 28 = ab(c - 1)(r - 1) + 1 + 3$

Another situation that arises in split plot designs wherein the treatments are not in a factorial arrangement is as follows. Suppose there are f families of genotypes with l lines selected from each of the families then the families are the whole plot treatments and the lines are the split plot treatments within each family. The whole plot treatments are replicated r times. A partitioning of the degrees of freedom in an analysis of variance is as follows:

Source of variation	Degrees of freedom
Total	rfl
Correction for the mean	1
Replicate = R	$r - 1$
Families = A	$f - 1$
Error $A = A \times R$	$(f - 1)(r - 1)$
Lines within families	$f(l - 1)$
Lines within family 1	$(l - 1)$
Lines within family 2	$(l - 1)$
Lines within family 3	$(l - 1)$
...	
Lines within family f	$(l - 1)$
Error B	$f(l - 1)(r - 1)$

Note that the number of lines within a family need not be equal.

3.5. SPLIT PLOT TREATMENTS IN AN INCOMPLETE BLOCK EXPERIMENT DESIGN WITHIN EACH WHOLE PLOT

For the second example discussed in Section 3.4, it may be desirable to use an incomplete block experiment design for the l lines of each family. Suppose that the incomplete block size is k and the incomplete block design is within each of the whole plot treatments (families) rather than within each replicate (complete block). There are b incomplete blocks of size k within a family and within each replicate, that is, $l = bk$. The partitioning of the degrees of freedom in an analysis of variance for each whole plot treatment (family) is

Family Source of variation	1 Degrees of freedom	2	3	...	f
Total	rl	rl	rl		rl
Correction for mean	1	1	1		1
Replicate = R	$r - 1$	$r - 1$	$r - 1$		$r - 1$
Lines = L	$l - 1$	$l - 1$	$l - 1$		$l - 1$
Blocks within R	$r(b - 1)$	$r(b - 1)$	$r(b - 1)$		$r(b - 1)$
Intrablock error	$rl - rb - l + 1$	$rl - rb - l + 1$	$rl - rb - l + 1$		$rl - rb - l + 1$

A combined analysis of variance would be as follows:

Source of variation	Degrees of freedom
Total	rfl
Correction for the mean	1
Replicate $= R$	$r - 1$
Families $= A$	$f - 1$
Error $A = A \times R$	$(f - 1)(r - 1)$
Lines within families	$f(l - 1)$
Blocks within replicates within families	$fr(b - 1)$
Intrablock error within families	$f(rl - rb - l + 1)$

Here again the number of lines in a family could vary. It would be a good idea to keep the incomplete block size k a constant or nearly so, in order to preserve the homogeneity of the intrablock errors. Furthermore, the incomplete block experiment design could be laid out within each complete block, but this destroys much of the orthogonality obtained with the previous arrangement.

3.6. SPLIT PLOT TREATMENTS IN A ROW-COLUMN ARRANGEMENT WITHIN EACH WHOLE PLOT TREATMENT AND IN DIFFERENT WHOLE PLOT TREATMENTS

Any appropriate experiment design may be used for the whole plot treatments. In some situations it may be desirable to design the split plot treatments in a row-column arrangement within a whole plot treatment. For example, suppose that there are five whole plots treatments, *A1, A2, A3, A4,* and *A5,* and four factor *B* split plot treatments, 1, 2, 3, and 4, in $r = 3$ complete blocks. A systematic arrangement of a row-column design for the split plot treatments might be as follows:

Replicate 1					Replicate 2					Replicate 3				
A1	*A2*	*A3*	*A4*	*A5*	*A1*	*A2*	*A3*	*A4*	*A5*	*A1*	*A2*	*A3*	*A4*	*A5*
1	1	1	1	1	2	2	2	2	2	3	3	3	3	3
2	2	2	2	2	3	3	3	3	3	4	4	4	4	4
3	3	3	3	3	4	4	4	4	4	1	1	1	1	1
4	4	4	4	4	1	1	1	1	1	2	2	2	2	2

Note that a four order (row) by three column Youden experiment design of columns and orders within columns, is used for the split plot treatments within each whole plot treatment. The whole plots are randomly allocated and then a separate randomization is used for the orders of each of the five whole plot treatments. An analysis of variance table presenting the partitioning of the degrees of freedom for each whole plot treatment would be:

	Whole plot treatment				
	A1	A2	A3	A4	A5
Source of variation		Degrees of freedom			
Total	12	12	12	12	12
Correction for mean	1	1	1	1	1
Replicate = R	2	2	2	2	2
Split plot treatments = B	3	3	3	3	3
Orders (eliminating B)	2	2	2	2	2
Error	4	4	4	4	4

A combined analysis of variance table partitioning of the degrees of freedom would be:

Source of variation	Degrees of freedom
Total	$60 = rab$
Correction for mean	1
Replicate = R	$2 = r - 1$
Whole plot treatments = A	$4 = a - 1$
$A \times R$ = Error A	$8 = (a - 1)(r - 1)$
B	$3 = b - 1$
$A \times B$	$12 = (a - 1)(b - 1)$
Orders within whole plots	$10 = a(o - 1)$
Residual = error within whole plots	$20 = a(b - 1)(r - 1) - a(o - 1)$

For the above arrangement, a balanced block arrangement of effects is maintained. If a Youden design of four rows and five columns had been used within each of the three replicates for factor B, an analysis of data from this experimental layout would still be possible, but precision would decrease due to confounding of effects. The balanced property for the Youden experiment design would still hold.

3.7. WHOLE PLOTS IN A SYSTEMATIC ARRANGEMENT

In some instances, the experimenter fails or is unable to randomize the whole plot treatments (Federer, 1955). Randomization is required for valid F-tests of the whole plot treatment effects. Hence in this case, there is no valid test of the whole plot treatment effects. Since the error mean square in systematic arrangements tends to be slightly over-estimated, the F-test will be on the 'conservative' side. The whole plot treatment effects are partially confounded with the whole plot treatment by replicate interaction effects.

3.8. SPLIT PLOTS IN A SYSTEMATIC ARRANGEMENT

In some instances, the experimenter fails to randomize the split plot treatments. One such example is described by Federer (1955), Table X-10. Nine fungicides, factor *A*, were arranged in a randomized complete block design with three replications. Three methods of applying the fungicides, factor *B*, were used. The experimenter stated that a split plot experiment design was used. In every one of the 27 whole plot experimental units, the same order of application was used. The net effect is that the design resembles a split block design with one set of treatments, methods of application, systematically arranged within each of the three replicates. It is not a split block design as the same order of methods of application were applied to *each* speu individually and not across all fungicides. A partitioning of the degrees of freedom in an analysis of variance table for this experiment is

Source of variation	Degrees of freedom
Total	$81 = rab$
Correction for the mean	1
Replicate $= R$	$2 = r - 1$
Fungicides $= A$	$8 = a - 1$
$A \times R =$ Error A	$16 = (a-1)(r-1)$
Methods of application $= B$	$2 = b - 1$
$A \times B$	$16 = (a-1)(b-1)$
Residual	$36 = a(b-1)(r-1)$

There is no valid test for method of application effects since *B* and the $A \times B$ interaction are partially confounded with Residual. The Residual mean square is biased upward by the systematic arrangement of methods of application.

3.9. CHARACTERS OR RESPONSES AS SPLIT PLOT TREATMENTS

Another variation of split plot designs is the situation wherein the produce is divided into different grades or when a mixture is divided into several parts. One example encountered during the course of statistical consulting, involved a randomized complete-block-designed experiment with factor *A* consisting of five fertilizer treatments within each of the five replicates (blocks) of a randomized complete block experiment design. The experiment was conducted to determine the treatment effect on yield and quality of strawberries. The experiment was actually laid out in five rows, the blocks, and five columns. The experimenter should have used a 5×5 Latin square experiment design. Even though he did not, the design is a five row by five column design because of the layout of the experiment. The berries in each of the 25 whole plot experimental units were graded into four quality grades, factor *B*, that were the split plot treatments. The response variables of interest were the weight and the number of strawberries in each of the four quality grades. A partitioning of the degrees of freedom in an analysis of variance table follows:

Source of variation	Degrees of freedom
Total	$100 = rab$
Correction for the mean	1
Replicate = row = R	$4 = r - 1$
Column = C	$4 = c - 1$
Treatment = A	$4 = a - 1$
Error A	$12 = (a-1)(r-1) - (c-1)$
Quality grades = B	$3 = b - 1$
$A \times B$	$12 = (a-1)(b-1)$
Error B	$60 = a(b-1)(r-1)$

The above is for a single picking at one calendar date. Actually, there were eight calendar picking dates for this experiment. A combined partitioning of the degrees of freedom in an analysis of variance table would be

Source of variation	Degrees of freedom
Total	$800 = abdr$
Correction for the mean	1
Dates	$7 = d - 1$
Replicates = rows within dates	$32 = d(r - 1)$
Rows	$4 = r - 1$
Rows × dates	$28 = (d-1)(r-1)$
Columns within dates	$32 = d(c - 1)$
Columns	$4 = c - 1$
Columns × dates	$28 = (c-1)(d-1)$
Treatments within dates	$32 = d(a - 1)$
Treatments = A	$4 = a - 1$
Treatments × dates	$28 = (a-1)(d-1)$
Error A within dates	$96 = d(a-1)(r-1)$
Quality grades within dates	$24 = d(b - 1)$
Quality grades = B	$3 = b - 1$
B × dates	$21 = (b-1)(d-1)$
$A \times B$ within dates	$96 = d(a-1)(b-1)$
$A \times B$	$12 = (a-1)(b-1)$
$A \times B \times$ dates	$84 = (a-1)(b-1)(d-1)$
Error B within dates = Error B	$480 = ad(b-1)(r-1)$

Since calendar dates are not replicated, there is no error to test for date main effects. The rows stay the same for all eight dates and thus the rows within date sum of squares may be partitioned into, that for rows and for rows by dates. The same comment holds for columns. The important picking date information for this type of experiment is the interaction of dates with factors A and B, on the number of strawberries and the weight of strawberries. The interaction of quality grade and the date of picking will be of interest. Other models for data of this type, involving the repeated nature of picking dates may be of interest to an analyst.

Another type of experiment that becomes a split plot design is when hay crops (factor A), say ten, are designed as a randomized complete block experiment design with six replicates, say then, the hay in each of the 60 whole plot experimental units is divided into weeds, legume, and grass species (factor B) to form the split plot treatments. A partitioning of the degrees of freedom in an analysis of variance table for this situation is

Source of variation	Degrees of freedom
Total	$180 = rab$
Correction for the mean	1
Replicate $= R$	$5 = r - 1$
Hay mixture $= A$	$9 = a - 1$
Error A $= A \times R$	$45 = (a - 1)(r - 1)$
Species $= B$	$2 = b - 1$
$A \times B$	$18 = (a - 1)(b - 1)$
Error B $= B \times R$ within A	$100 = a(b - 1)(r - 1)$

The hay mixture effects and the $A \times B$ interaction effects would be the items of interest rather than a test of species effect, the factor B effect.

3.10. OBSERVATIONAL OR EXPERIMENTAL ERROR?

A split plot design was used in a poultry breeding experiment to study the length of fertility of sperm in the oviduct of two strains of chicken (Federer, 1955, Table X-11). Two pens representing replicates were used. The whole plot treatments were eight cocks, four from one strain and four from the other. The eight cocks were placed in both pens. Each cock was mated with four hens, two hens from one strain and two hens from the other. This resulted in 32 hens in each of the pens. A breakdown of the degrees of freedom in an analysis of variance for this experiment is as follows:

Source of variation	Degrees of freedom
Total	$64 = hnpc$
Correction for the mean	1
Pens	$1 = p - 1$
Cocks	$7 = c - 1$
Between strains	$1 = s - 1$
Within strain one	$3 = c/2 - 1$
Within strain two	$3 = c/2 - 1$
Error A $=$ pens $\times$ cocks	$7 = (c - 1)(p - 1)$
Between strains for hens	$1 = h - 1$
Strains for hens $\times$ strains for cocks	$1 = (h - 1)(s - 1)$
Error B	$14 = s(h - 1)(c - 1)$
Between two strains, one hen on same cock	$16 = hp(n - 1)(c/2)$
Between two strains, two hens on same cock	$16 = hp(n - 1)(c/2)$

A few data analysts may consider pooling the last two sums of squares with 32 degrees of freedom and use this as the error term instead of Error B. This would be incorrect as the component of variance for hens × pens is not included in this mean square which is observational or sampling error. This type of error occurs frequently as analysts use sampling or observational error mean squares when the experimental error mean square is required. The hens × pens variance component is included in Error B, between strains for hens, and the interaction of strains hens × strains cocks mean squares. Hence, Error B is the correct error mean square to use in testing for the hen strain effect and for the hen strain × cock strain interaction effect.

3.11. TIME AS A DISCRETE FACTOR RATHER THAN AS A CONTINUOUS FACTOR

A number of investigations involve the recording of response variables over time. Two ways of handling the time factor is first to compute trend, curvature of trend, and so forth, with time as the independent continuous factor, and secondly to treat time as discrete levels of a factor. The first method would allow calculation of interactions both with trend and with curvature. The latter method allows for interactions of time periods with other factors in the investigation. This latter use is the consideration of this section. Investigations on hay crops involve several cuttings over a season. Experiments involving vegetables often have several pickings or harvestings. The decision related to the treatment of time levels in an experiment, is determined by what determines the time periods. Many investigations set the time at specific calendar dates such as June 15, July 15, August 15, and so forth, that is, the time intervals are *calendar clock times*. For such investigations, there is no replication of time periods or dates. Suppose a randomized complete-block-designed experiment with b hay crops in r replicates is harvested on a calendar dates over a season, a partitioning of the degrees of freedom in an analysis of variance table is as follows:

Source of variation	Degrees of freedom
Total	arb
Correction for the mean	1
Time levels	$a - 1$
Replicates or complete blocks	$r - 1$
Time × replicate	$(a - 1)(r - 1)$
Hay crops	$b - 1$
Time × hay crop	$(b - 1)(a - 1)$
Hay crop × replicate within time	$a(b - 1)(r - 1)$

Because the calendar time levels are un-replicated, there is no error term to test for time period effects. Ordinarily such a test would not be desired. To test for effects of hay crop and hay crop × time interaction, the hay crop × replicates within time period mean square is the appropriate error term.

Now consider the same setup but let the time intervals be determined for each experimental unit of the factors replicate and hay crop. For example, if a legume is present in the hay crop, the time to harvest an experimental unit is determined by when three-fourths, say, of the legume plants are in bloom. Thus, a *biological clock time* is used to determine when to harvest a particular hay-crop-replicate experimental unit. This has the effect of randomizing time levels, biological clock times, within each of these experimental units. The effect of this procedure is illustrated in the following partitioning of the degrees of freedom in an analysis of variance:

Source of variation	Degrees of freedom
Total	rab
Correction for the mean	1
Replicate or complete block	$r - 1$
Hay crop	$b - 1$
Hay crop × replicate	$(b - 1)(r - 1)$
Time	$a - 1$
Hay crop × time	$(a - 1)(b - 1)$
Time × replicate within hay crop	$b(r - 1)(a - 1)$

The hay crop × replicate mean square is used to test for hay crop effects. The time × replicate within hay crop mean square is used to test for the presence of time and hay crop × time interaction effects. For this arrangement, biological clock times become the split plot treatments whereas calendar times are the whole plot treatments in the previous arrangement. The distinction between calendar clock times and biological clock times illustrates one of the many subtleties encountered in designing split plot experiments.

To illustrate analyses when time is involved in a split split plot experiment design, the following example was encountered during a statistical consulting session. Three methods of cultivation for potato plants were arranged in a randomized complete block experiment design with $r = 4$ blocks. Methods of cultivation were the whole plot treatments. Four varieties of potatoes represented the split plot treatments. The split plot experimental unit consisted of three rows of potato plants with two feet between plants in a row. The response variable was weed count at three stages of growth of the plants and the counts were made in the spaces between the plants. Stage 1 was when the potato plants were "touching". Stage 2 was when the plants began to vine, "vining". Stage 3 was when the plants were "mature". The stages were determined for the experiment, and not for each individual speu. Thus, "clock time" or "calendar time" was used. "Biological clock time" would have been used if the stage was determined for each speu. A standard split plot design analysis could be performed for the weed counts at each of the three stages. Also, this analysis could be performed on the differences in weed counts at two stages, say stage 2 – stage 1. A partitioning of the total degrees of freedom for the experiment in an analysis of variance table is as follows:

Source of variation	Degrees of freedom
Total	$144 = bcsv$
Correction for the mean	1
Blocks $= B$	$3 = b - 1$
Stages $= S$	$2 = s - 1$
$B \times S =$ Error S	$4 = (b-1)(s-1)$
Cultivation method $= C$	$2 = c - 1$
$C \times S$	$4 = (c-1)(s-1)$
$B \times C$ within $S =$ Error C	$18 = s(b-1)(c-1)$
Varieties $= V$	$3 = v - 1$
$V \times S$	$6 = (s-1)(v-1)$
$V \times C$	$6 = (c-1)(v-1)$
$V \times C \times S$	$12 = (c-1)(s-1)(v-1)$
$V \times B$ within S and $C =$ Error V	$81 = cs(b-1)(v-1)$

Note that if "biological clock time" had been used, then stage would have been the split split plot.

The above examples and the following example were obtained when a researcher required statistical assistance for the analysis of data from an experiment conducted as described. An experiment on pheasants was conducted according to the following layout:

Treatment 1			Treatment 2			Treatment 3		
	Method			Method			Method	
	I	II		I	II		I	II
Bird 1			Bird 5			Bird 9		
Stage 1	x	x	Stage 1	x	x	Stage 1	x	x
Stage 2	x	x	Stage 2	x	x	Stage 2	x	x
Stage 3	x	x	Stage 3	x	x	Stage 3	x	x
Stage 4	x	x	Stage 4	x	x	Stage 4	x	x
Mature	x	x	Mature	x	x	Mature	x	x
Bird 2			Bird 6			Bird 10		
Stage 1	x	x	Stage 1	x	x	Stage 1	x	x
Stage 2	x	x	Stage 2	x	x	Stage 2	x	x
Stage 3	x	x	Stage 3	x	x	Stage 3	x	x
Stage 4	x	x	Stage 4	x	x	Stage 4	x	x
Mature	x	x	Mature	x	x	Mature	x	x
Bird 3			Bird 7			Bird 11		
Mature	x	x	Mature	x	x	Mature	x	x
Bird 4			Bird 8			Bird 12		
Mature	x	x	Mature	x	x	Mature	x	x

The symbol x denotes where an observation was obtained. The stage was determined for each of the 12 birds, that is, "biological clock time" was used. Four birds were randomly assigned to each treatment. Methods I and II were used in a random order for each stage and each bird. To obtain a partitioning of the degrees of freedom, first consider only the responses for birds 1, 2, 5, 6, 9, and 10, as the data for all five stages were obtained for these birds. This results in the following partitioning of the total degrees of freedom:

Source of variation	Degrees of freedom
Total	$60 = bmst$
Correction for the mean	1
Treatments $= T$	$2 = t - 1$
Birds within $T =$ Error T	$3 = t(b - 1)$
Stages $= S$	$4 = s - 1$
$S \times T$	$8 = (s - 1)(t - 1)$
$S \times B$ within $T =$ Error S	$12 = t(b - 1)(s - 1)$
Method $= M$	$1 = m - 1$
$M \times T$	$2 = (m - 1)(t - 1)$
$M \times S$	$4 = (m - 1)(s - 1)$
$M \times T \times S$	$8 = (m - 1)(t - 1)(s - 1)$
$M \times$ Birds within T and $S =$ Error M	$15 = st(b - 1)(m - 1)$

The additional 12 degrees of freedom from Birds 3, 4, 7, 8, 11, and 12 go into an analysis of variance table as follows:

Source of variation	Degrees of freedom
Total	72
Correction for the mean	1
Treatments $= T$	2
Birds within $T =$ Error T	9
Stages $= S$	4
$S \times T$	8
$S \times B$ within $T =$ Error S	12
Method $= M$	1
$M \times T$	2
$M \times S$	4
$M \times T \times S$	8
$M \times$ Birds within T and $S =$ Error M	21

Birds within treatments now have nine instead of three degrees of freedom. The other six degrees of freedom end up in Error M.

The following is a description of an experiment conducted on fish. The goal of the experiment was to determine if fish could detect an odor, that is, do they have a sense of smell. The fish were from two stocks but the stocks were of different ages and therefore stock and age effects were completely confounded. The fish of one of the stocks used in this experiment were originally grown in an environment where the odor was present. The fish of the other stock were grown in water without the odor. A container with four un-partitioned chambers in a two by two arrangement was used. Two water streams entered one side of the container, flowing through the container, and out from the opposite side. One of the water streams contained the odor and the other did not. Thus, the whole plot treatments were in a two by two factorial arrangement of the two factors, odor and stock-age, that is, $t = 4$. The split plot treatments were the four chambers. Twenty runs of the experiment were conducted. The whole plot treatments were in a completely randomized design with an unequal number of fish per whole plot treatment. Around seven to eleven fish were used in each run. The response measured was the number of fish observed in each of the four chambers, $c = 4$, at any given time. Counts were made every minute for 250 minutes. The odor was introduced into the water so the heaviest concentration of the odor was in that chamber. The next chamber beyond and toward the outlet had considerably less of the odor. The chamber, where the water without any odor was introduced, had no odor and the next chamber beyond that, next to the outlet, had very little of the odor that diffused into it.

One question that arises is why the chambers are considered to be split plot treatments. Note that there were $20 = cr$ runs. For each run, a different group of fish and one combination of the factors, odor and stock-age, was used to obtain the responses of the number of fish in any given chamber. In general, it is always essential to understand the exact and precise way in which an experiment was conducted in order to determine the appropriate statistical analysis.

The experimenter stated that she wanted to know the results at the end of the 250 observations taken at one minute intervals. However, there is additional information in the data. The first item to be considered is what to do with the 80 sets of 250 minute observations. A plot of the number of fish in a chamber against time for some of these 80 sets should indicate whether fish were randomly distributing themselves in a chamber over time or whether a point of stability in number of fish in a chamber was reached. It was thought that fish would congregate in the chamber with the most odor if they had a sense of being able to detect an odor. This means that the number of fish in a chamber should stabilize at some point in time, that is, the response curve should reach an asymptote. If interest centered on the nature of the response curve over time, then some appropriate model should be selected and the parameters of this model estimated in a statistical analysis.

Alternatively, the average number of fish in five 50 minutes time periods, say, could be used. Then, the $r = 5$ time periods could be used as split split plot treatments. On the contrary, a statistical analysis could be made for an average of the responses from minutes 201 to 250, or any other time interval. The partitioning of the degrees of freedom for one such set of observations could be as follows:

TIME AS A DISCRETE FACTOR RATHER THAN AS A CONTINUOUS FACTOR

Source of variation	Degrees of freedom
Total	$80 = crt$
Correction for mean	1
Whole plot treatments	$3 = t - 1$
Odor	$1 = o - 1$
Stock-age	$1 = s - 1$
Odor × stock-age	$1 = (o - 1)(s - 1)$
Runs within odor and stock-age	$16 = os(r - 1)$
Chamber	$3 = c - 1$
Chamber × odor	$3 = (c - 1)(o - 1)$
Chamber × stock-age	$3 = (c - 1)(s - 1)$
Chamber × odor × stock-age	$3 = (c - 1)(o - 1)(s - 1)$
Error	$48 = t(c - 1)(r - 1)$

The "runs within odor and stock-age" degrees of freedom and sum of squares may be partitioned into runs within each of the four whole plot treatments. The "Error" degrees of freedom and sum of squares may be partitioned into chambers × runs within each of the four whole plot treatments.

The question now arises as to what to do with the response "number of fish in a chamber". Remember that the number of fish used in any run varied from seven to eleven. One could use the proportion of number of fish in a chamber and perhaps an arcsine or a square root transformation of the proportions. Alternatively, one could use number of fish used in a run as a covariate. Neither of these procedures will completely take care of the problem of unequal numbers of fish, but they will alleviate it.

If five, say, time classes were used, a partitioning of the degrees of freedom in an analysis of variance table would be as follows:

Source of variation	Degrees of freedom
Total	$400 = cprt$
Correction for mean	1
Whole plot treatments	$3 = t - 1$
Odor	$1 = o - 1$
Stock-age	$1 = s - 1$
Odor × stock-age	$1 = (o - 1)(s - 1)$
Runs within odor and stock-age	$16 = t(r - 1)$
Chamber	$3 = c - 1$
Chamber × odor	$3 = (c - 1)(o - 1)$
Chamber × stock-age	$3 = (c - 1)(s - 1)$
Chamber × odor × stock-age	$3 = (c - 1)(o - 1)(s - 1)$
Error chamber	$48 = t(c - 1)(r - 1)$
Time periods	$4 = p - 1$
Time × odor	$4 = (p - 1)(o - 1)$

(*Continued*)

(Continued)

Source of variation	Degrees of freedom
Time × stock-age	$4 = (p-1)(s-1)$
Time × odor × stock-age	$4 = (p-1)(o-1)(s-1)$
Time × chamber	$12 = (c-1)(p-1)$
Time × chamber × odor	$12 = (c-1)(o-1)(p-1)$
Time × chamber × stock-age	$12 = (c-1)(s-1)(p-1)$
Time × chamber × odor × stock-age	$12 = (c-1)(o-1)(s-1)(p-1)$
Error time	$256 = ct(r-1)(t-1)$

Instead of using five time periods, the nature of the response curve may indicate another partitioning of the 250 minute responses. For example, it may be found that for a given whole plot treatment, the number of fish in a chamber becomes stable at t minute. Then two time classes could be used, the average number from one min up to the point of the stability minute, t, and the average of the remaining responses from $t + 1$ minute up to the 250 minute mark. It is worth mentioning that the point of stability may vary from whole plot to whole plot treatment.

3.12. INAPPROPRIATE MODEL?

A split plot designed experiment involving three varieties of oats, the whole plots v1, v2, and v3, and four manurial treatments, the split plots n0, n1, n2, and n3, in six replicates, r1 to r6, is described by Yates (1937). The plan of the *presumed* layout of the experiment and the yields in 1/4 pounds from the experiment are given in Table 3.1 and a standard analysis of variance is given in Table 3.2. It is stated that this is the plan and it is *assumed to be* the layout in the field. The layout appears to be an eighteen row by four column arrangement. The split plot sizes were 1/80 of an acre and this size should be large enough to minimize competition between split plot experimental units.

There are two things that are bothersome about the standard split plot model. Note that the coefficient of variation for split plot units is 12.8% that is rather high for oats experiments. Secondly, the variety by nitrogen mean square is much smaller than the nitrogen × replicate within variety mean square. Using the greater mean square rule as described by Robson (1953), we compute $F = \frac{177.08}{53.63} = 3.30$. This would indicate significance at about the 12% level. Perhaps other models would be appropriate.

An alternate model could be to use logarithms of the yields as shown in Table 3.3. The log transformation of the data reduced the coefficient of variation, but did little to improve the discrepancy between the variety-nitrogen interaction and the error mean square. The F-value was increased to 0.40 from 0.30. The square root transformation did less than the log transformation in accounting for the items of concern about this experiment.

Table 3.1. Yields in 1/4 Pounds for an Oat Variety, Whole Plot, and Manurial (Nitrogen), Split Plot, and Trial Designed as a Split Plot Experiment Design in Six Complete Blocks (Replicates).

r1	v3	n3	156	n2	118	n2	109	n3	99	v3	r4
		n1	140	n0	105	n0	63	n1	70		
	v1	n0	111	n1	130	n0	80	n2	94	v2	
		n3	174	n2	157	n3	126	n1	82		
	v2	n0	117	n1	114	n1	90	n2	100	v1	
		n2	161	n3	141	n3	116	n0	62		
r2	v3	n2	104	n0	70	n3	96	n0	60	v2	r5
		n1	89	n3	117	n2	89	n1	102		
	v1	n3	122	n0	74	n2	112	n3	86	v1	
		n1	89	n2	81	n0	68	n1	64		
	v2	n1	103	n0	64	n2	132	n3	124	v3	
		n2	132	n3	133	n1	129	n0	89		
r3	v2	n1	108	n2	126	n2	118	n0	53	v1	r6
		n3	149	n0	70	n3	113	n1	74		
	v3	n3	144	n1	124	n3	104	n2	86	v2	
		n2	121	n0	96	n0	89	n1	82		
	v1	n0	61	n3	100	n0	97	n1	99	v3	
		n1	91	n2	97	n2	119	n3	121		

Table 3.2. Analysis of Variance for Standard Split Plot Experiment Design and F-Tests for Data of Table 3.1.

Source of variation	Degrees of freedom	Sum of squares	Mean square	F-value
Total	72	830,322		
Correction for mean	1	778,336.06		
Replicate	5	15,875.28	3175.06	
Variety	2	1786.36	893.18	1.49
Replicate × variety	10	6013.31	601.33	
Nitrogen (manure)	3	20,020.50	6673.50	37.69
Variety × nitrogen	6	321.75	53.63	0.30
Rep. × nit. within var.	45	7968.75	177.08	

Table 3.3. Analysis of Variance and F-Values for Log(Yield) Values of the Data in Table 3.1.

Source of variation	Degrees of freedom	Sum of squares	Mean square	F-value
Total (corrected)	71	5.1458		
Replicate	5	1.3318	0.2664	
Variety	2	0.2274	0.1137	2.01
Variety × replicate	10	0.5661	0.0566	
Nitrogen	3	2.1351	0.7117	38.11
Variety × nitrogen	6	0.0451	0.0075	0.40
Rep. × nit. within var.	45	0.8404	0.0187	

Table 3.4. Another Model for the Data of Table 3.1.

Source of variation	Degrees of freedom	Type I Mean square	Degrees of freedom	Type III Mean square	F-value
Replicate	5	3175.06	2	2971.26	
Variety	2	893.18	2	1399.90	5.61
Replicate × variety	10	601.33	4	249.45	
Nitrogen	3	6673.50	3	3992.62	34.84
Nitrogen × variety	6	53.63	6	79.64	0.69
Row	9	100.10	9	105.00	
Column	2	1585.75	2	1585.75	13.84
Residual	34	114.60	34	114.60	

A model taking into account the 18-row by 4-column layout of the experiment should do better than the above models in accounting for the spatial variation present, Table 3.4. A possible model in SAS POC GLM format is

Yield = replicate variety replicate × variety nitrogen variety × nitrogen row column.

There will be a high degree of confounding of effects in this model and hence a Type III analysis will be appropriate. For example, the contrast columns 1 + 2 versus columns 3 + 4 will be completely confounded with replicate effect. This accounts for the fact that the Type III analysis for columns indicates only 2 rather than 3 degrees of freedom for columns. The coefficient of variation is now 10.3%, but still rather high. The F-value for the variety by nitrogen interaction has increased to 0.69 which is approaching the expected value of one, if the null hypothesis of zero interaction is true. Using a number of other models, as suggested by Federer (2003), did not improve on the above model. Note that the above model indicates a significant difference among varieties at about the 7% level. $F_{.10}(2,4) = 4.32$ and $F_{.05}(2,4) = 6.94$. It may be possible that the plan given by Yates (1937) is not the actual layout of the experiment and this could be the reason why a more appropriate model could not be found.

It may be that there is variation in the variety by replicate interaction that should be taken into account. A linear replicate by linear variety effect was somewhat helpful in reducing this mean square from 601.33 to 513.80, Table 3.5. A study of the variety by replicate display in Figure 3.1 indicates great disparity of varietal yields from replicate to replicate. This appears to be unsuspected since the variety whole plots are 1/20 acre in size.

Table 3.5. Another Model for the Whole Plot Analysis for the Data of Table 3.1.

Source of variation	Degrees of freedom	Sum of squares	Mean square	F-value
Replicate	5	15,875.28	3175.06	
Variety	2	1786.36	893.18	1.74
Non-additivity	1	1389.15	1385.15	2.70
Residual	9	4624.16	513.80	

INAPPROPRIATE MODEL?

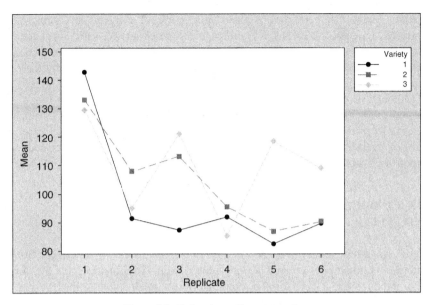

Figure 3.1. Variety by replicate interaction.

A study of the display in Figure 3.1 indicates that there are apparent outliers in replicates 3, 5, and perhaps 6. At this point the analyst of these data should contact the experimenter to determine what happened to make the variety yields fluctuate so wildly from replicate to replicate. This is in contrast to the display in Figure 3.2,

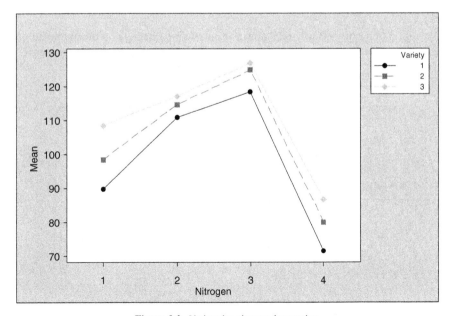

Figure 3.2. Variety by nitrogen interaction.

which is an almost perfect zero interaction graph. If a mean rank sum test is conducted on these 12 means, a significant difference among varieties at the 2% level is indicated (See, e.g., Federer, 2004). The intercepts are different and this, rather than trends, may account for the apparent interaction indicated by the rank sum test.

Although models that improve the analysis of these data have been found, the original problems are not completely resolved. Discussions with the experimenter are in order prior to drawing any conclusions about the effects, other than about the nitrogen treatments, or for investigating other models. A SAS computer code and output of the code are given in Appendix 3.1.

3.13. COMPLETE CONFOUNDING OF SOME EFFECTS AND SPLIT PLOT EXPERIMENT DESIGNS

There is a direct relationship between complete confounding of effects from a factorial treatment design and a split plot design. To illustrate, let the three-factor interaction in a 2^3 factorial be completely confounded, that is, the levels of the three-factor interaction ABC_0 and ABC_1 become whole plot treatments (Federer, 1955). A systematic plan is as follows:

Replicate	1		2		3		...	r	
	ABC_0	ABC_1	ABC_0	ABC_1	ABC_0	ABC_1		ABC_0	ABC_1-
	000	100	000	100	000	100		000	100
	110	010	110	010	011	010		110	010
	101	001	101	001	101	001		101	001
	011	111	011	111	011	111		011	111

The split plot treatments are the factorial combinations making up the combinations in ABC_0 and ABC_1. A key-out of the degrees of freedom in an analysis of variance is as follows:

Source of variation	Degrees of freedom
Total	$8r$
Correction for mean	1
Replicate	$r-1$
$A \times B \times C$	1
Replicate $\times A \times B \times C$ (Error A)	$r-1$
A	1
B	1
C	1
$A \times B$	1
$A \times C$	1
$B \times C$	1
Error B	$6(r-1)$

The completely confounded effects take on the role of whole plot treatments, and the combinations making up the levels of these effects are the split plot treatments. Further discussion of this type of design may be found in Federer (1955) and Singh (1950).

Bingham *et al.* (2004) have developed a systematic method to generate treatment designs that have a fractional factorial design structure in split plot arrangements.

3.14. COMMENTS

Although many variations of a split plot designed experiment are described in this chapter, there are many more that arise in the course of experimentation and in statistical consulting. The number and variety presented herein should offer a guide toward handling other situations. As in the examples given in previous sections, the design is not always obvious but requires some thought. Spatially designed experiments are easier to comprehend than those found in animal, human, educational, ecological, and other experiments. In some of the statistics courses, split plot designed experiments are treated as if everything fits into the Chapter 1 framework. Consideration of all types of split plot designed experiments indicate that the philosophy and theoretical considerations are not trivial. Further discussion of types of design variation that occurs in split plot designed experiments may be found in Federer (1975, 1977).

For the first example in Section 3.9, computer software does not readily give the row-column error mean square, Error A. One way to obtain this error mean square is to use the following two SAS/GLM MODEL statements:

$$\text{Response} = \text{row column } A\ B\ A^*B;$$
$$\text{Response} = \text{column } A\ A^*\text{column } B\ A^*B$$

The Error A term may be obtained as the difference between A*column and row sums of squares or between the residual mean squares for the first model and the second one.

In using SAS software for analyses involving confounding, it has been found that the degrees of freedom and sums of squares may not be what is expected. Several situations have been found where one expects one thing from the model used, but SAS output does not agree with what is known to be true.

3.15. PROBLEMS

Problem 3.1. Obtain the residuals for the models described in Tables 3.2 to 3.5. Obtain stem and leaf plots to determine outliers and/or patterns.

Problem 3.2. A seed germination experiment on $a = 49$ strains of guayule with $b = 4$ seed treatments in $r = 6$ replicates was conducted to determine the germination percentage and the average number of days before plant emergence following seeding (Federer, 1946). The original experiment design was two repetitions of a triple lattice incomplete block experiment design for the 49 guayule strains with split plots of size four for the $b = 4$ seed treatments. The incomplete block design was for $a = v = 49$, $r = 6$ with incomplete blocks of size $k = 7$. The original data were unavailable. Using the actual means and standard errors given by Federer (1946), data for average number of days to emergence were simulated for the experiment. The data for replicates 1 and 2 only are given on the book's FTP site, as the data for all six replicates was considered to be too lengthy for a problem. Data for all six replicates are available upon request. Such an experiment of this size is not unusual in plant breeding investigations. Many are much larger. Obtain an analysis of the data for both intrablock analysis and for recovery of interblock information analysis. Also, write a code for ranking the strain means and the strain*treatment means from earliest time of emergence to latest.

```
/*row.names Replicate Strain Treatment Block TimeDays*/

1    Rep1   S1    T1    Block5   6.01
2    Rep1   S1    T2    Block5   7.16
3    Rep1   S1    T3    Block5   8.65
4    Rep1   S1    T4    Block5   7.45
5    Rep1   S2    T1    Block2   5.91
6    Rep1   S2    T2    Block2   7.17
7    Rep1   S2    T3    Block2   7.7

....

389  Rep2   S49   T1    Block4   6.36
390  Rep2   S49   T2    Block4   9.06
391  Rep2   S49   T3    Block4   6.49
392  Rep2   S49   T4    Block4   11.5   ;
```

Problem 3.3. A split split plot design was used to investigate the effect of the following factors on planting method (main/whole plot factor, P: 3 levels), stage of fertilizer application (split plot factor, S: 3 levels), and on the yield of soybean varieties (split split plot factor, V: 7 levels). The design layout on the main plots was a randomized complete block design (Block, R: 4 levels). The layout for this split split plot design was done as follows:

Step #1: The experimental field was divided into four blocks. Each block (18 m by 14 m each) was divided into 3 plots to which the 3 levels of factor P were randomly assigned. Thus P was the whole plot factor. The whole plot experimental unit per level of P was 6 m by 14 m.

Step #2: Within each block, each whole plot experimental unit was further divided into 3 plots to which the three levels of factor S were randomly assigned. Thus S was the split plot factor. The split plot experimental unit per level of S was 2 m by 14 m.

Step #3: Finally, within each block, each split plot experimental unit for each level of S was further divided into seven split split plots to which the seven levels of factor V

were randomly assigned. Thus V was the split split plot factor. The split split plot experimental unit was 2 m by 2 m.

The data for this problem given on the accompanying disk are as follows (four significant figures are ample for statistical analyses, i.e., the two decimal digits may be dropped.):

(i) Give a linear model for analyzing the data for this design, and show the required statistical assumptions.

(ii) Conduct an analysis of variance for the data in this study using SAS PROC GLM and give an interpretation of your results.

(iii) Estimate the variance of the mean for the ith level of factor P (planting method), $\bar{y}_{i..}$.

(iv) Give a formula for the variance of the difference between two means $\bar{y}_{ij.}$ and $\bar{y}_{.lm.}$, i.e. $\text{var}(\bar{y}_{ij.} - \bar{y}_{.lm.})$, where i and l are the levels of the whole plot factor P and j and m are levels of the split plot factor S.

(v) Estimate the standard error for the difference between two variety means.

(vi) What is the standard error of the difference between two planting method means?

(viii) What is the standard error of the difference between two stage of fertilizer application means?

(ix) Find a 95% confidence interval for the difference between two means of factor P.

(x) Estimate the variance of the mean for one level of factor S.

(xi) Estimate the standard error of the difference between two means $\bar{y}_{ij.}$ and $\bar{y}_{.lm.}$, that is $\text{var}(\bar{y}_{ij.} - \bar{y}_{.lm.})$, where i and l are the levels of the whole plot factor P and j and m are levels of the split plot factor S.

3.16. REFERENCES

Bingham, D. R., E. D. Schoen, and R. R. Sitter (2004). Designing fractional factorial split-plot experiments with few whole-plot factors. Applied Statistics 53(2):325–339.

Federer, W. T. (1946). Variability of certain seed, seeding, and young-plant characters of guayule. USDA, Technical Bulletin No. 919, August, Washington, D. C.

Federer (1955). *Experimental Design: Theory and Application.* Macmillan, New York (Reprinted by Oxford and IBH Publishing Company, New Delhi, 1962 and 1974), Chapter X, pp. 271–306.

Federer, W. T. (1975). The misunderstood split plot. In *Applied Statistics* (Proceedings of a Conference at Dalhousie University, Halifax, Nova Scotia, May 2–4, 1974; Editor: R. P. Gupta), North-Holland Publishing Company, Amsterdam, Oxford, pp. 145–153.

Federer, W. T. (1977). Sampling, model, and blocking considerations for split plot and split block designs. Biometrische Zeitschrift (Biometrical Journal) 19:181–200.

Federer, W. T. (2003). Exploratory model selection for spatially designed experiments–Some examples. Journal of Data Science 1(3):231–248.

Federer, W. T. (2004). Non-parametric procedures for comparing a set of response curves. BU-1655-M in the Technical Report Series of the Department of Biological Statistics and Computational Biology, Cornell University, Ithaca, NY 14853, July.

Robson, D. S. (1953). A common misconception concerning the "greater mean square rule". BU-38-M in the Technical Report Series of the Biometrics Unit (now the Department of Biological Statistics and Computational Biology), Cornell University, Ithaca, New York 14853.

Singh, M. (1950). Confounding in split-plot designs with restricted randomization of sub-plot treatments. Empire Journal of Agriculture 18:190–202.

Yates, F. (1937). The design and analysis of factorial experiments. Technical Communication No. 35, Imperial Bureau of Soil Science, Harpenden, England.

APPENDIX 3.1. TABLE 3.1 CODE AND DATA

The data for Table 3.1 are given on the book's FTP site.

```
Data spexvar3;
Input row col rep var nit set reps yield;
Datalines;

1   1   1   3   3   1   1   156
1   2   1   3   2   1   1   118
1   3   4   3   2   2   1   109

....

18  1   3   1   1   1   3   91
18  2   3   1   2   1   3   97
18  3   6   3   2   2   3   119
18  4   6   3   3   2   3   121
;

Proc GLM data = spexvar3;

Class rep var nit;
Model yield = rep var rep*var nit var*nit;
lsmeans var nit var*nit;

run;

Proc GLM data = spexvar3;
Class rep var nit row col;
Model yield = rep var rep*var nit var*nit row col;
run;
```

An abbreviated form of the output from running the above code and data set is presented below:

APPENDIX

The GLM Procedure

Dependent Variable: yield

Source	DF	Sum of Squares	Mean Square	F Value	Pr > F
Model	26	44017.19444	1692.96902	9.56	<.0001
Error	45	7968.75000	177.08333		
Corrected Total	71	51985.94444			

R-Square	Coeff Var	Root MSE	yield Mean
0.846713	12.79887	13.30727	103.9722

Source	DF	Type I SS	Mean Square	F Value	Pr > F
rep	5	15875.27778	3175.05556	17.93	<.0001
var	2	1786.36111	893.18056	5.04	0.0106
rep*var	10	6013.30556	601.33056	3.40	0.0023
nit	3	20020.50000	6673.50000	37.69	<.0001
var*nit	6	321.75000	53.62500	0.30	0.9322

Least Squares Means

var	yield LSMEAN
1	97.625000
2	104.500000
3	109.791667

nit	yield LSMEAN
1	98.888889
2	114.222222
3	123.388889
4	79.388889

var	nit	yield LSMEAN
1	1	89.666667
1	2	110.833333
1	3	118.500000
1	4	71.500000
2	1	98.500000
2	2	114.666667
2	3	124.833333
2	4	80.000000
3	1	108.500000
3	2	117.166667
3	3	126.833333
3	4	86.666667

Dependent Variable: yield

Source	DF	Sum of Squares	Mean Square	F Value	Pr > F
Model	37	48089.54984	1299.71756	11.34	<.0001
Error	34	3896.39461	114.59984		
Corrected Total	71	51985.94444			

R-Square	Coeff Var	Root MSE	yield Mean
0.925049	10.29615	10.70513	103.9722

Source	DF	Type I SS	Mean Square	F Value	Pr > F
rep	5	15875.27778	3175.05556	27.71	<.0001
var	2	1786.36111	893.18056	7.79	0.0016
rep*var	10	6013.30556	601.33056	5.25	0.0001
nit	3	20020.50000	6673.50000	58.23	<.0001
var*nit	6	321.75000	53.62500	0.47	0.8271
row	9	900.85852	100.09539	0.87	0.5576
col	2	3171.49687	1585.74844	13.84	<.0001

Source	DF	Type III SS	Mean Square	F Value	Pr > F
rep	2	5942.52778	2971.26389	25.93	<.0001
var	2	2799.80000	1399.90000	12.22	0.0001
rep*var	4	997.78333	249.44583	2.18	0.0926
nit	3	11977.85911	3992.61970	34.84	<.0001
var*nit	6	477.82134	79.63689	0.69	0.6553
row	9	945.00685	105.00076	0.92	0.5230
col	2	3171.49687	1585.74844	13.84	<.0001

CHAPTER 4

Variations of the Split Block Experiment Design

4.1. INTRODUCTION

There are many ways to vary a split block experiment design, just as there are for the split plot experiment design. These variations arise owing to the need or desire of the experimenter to conduct an experiment in a certain manner. For the statistical consultant, it is always wise to have the experimenter describe *in detail* how the experiment was laid out and conducted. A clear understanding of this allows one to present an appropriate statistical analysis for the data from the experiment. It is not sufficient to ask the experimenter to name the design that was thought to have been used. The conduct and layout of the experiment are important factors in determining the actual design of the experiment. Several variations of the split block experiment design are described in the following sections.

In Section 4.2, a statistical analysis is presented for a split-block-designed experiment with one set of treatments, factor A, in a randomized complete block design and the second set of treatments, factor B, in a row-column design. In Section 4.3, both of the factors are designed as row-column experiment designs. In Section 4.4, a *split block split-block*-designed experiment for three factors is described. Factor A with five treatments is designed as a 5×5 Latin square design. A second set of four treatments, factor B, uses the columns as the blocks of a randomized complete block design. The third set of treatments, factor C, is in a randomized complete block design with the rows as blocks for this design. A numerical example is used to describe the statistical analysis of the design and the SAS code for the analysis provided.

Variations on Split Plot and Split Block Experiment Designs, by Walter T. Federer and Freedom King
Copyright © 2007 John Wiley & Sons, Inc.

The analysis of a split block design with one set of treatments in an incomplete block design and the second set in a randomized complete block design is presented in Section 4.5. In Section 4.6, a statistical analysis is described for the arrangement where one set of treatments is designed in a randomized complete block design and the second set of treatments is laid out over all treatments of factor A and over all blocks. That is, there is no replication on the second set of treatments and only one arrangement or randomization. The concept of confounding interactions of a factorial treatment arrangement as the whole plot treatments is discussed in Section 4.7. In Section 4.8, designs with one or more controls common to both factors are discussed. The examples and illustrations described herein arose during the course of offering statistical advice to an experimenter. Finally, some general comments are presented in Section 4.9.

4.2. ONE SET OF TREATMENTS IN A RANDOMIZED COMPLETE BLOCK AND THE OTHER IN A LATIN SQUARE EXPERIMENT DESIGN

Consider the split block experiment design where one set of treatment, factor A, is designed as a Latin square and a second set of treatments, factor B, is designed as a randomized complete block with either the rows or the columns being the blocks (See Example 2.2). An experiment designed in this fashion had five rootstocks, R1, R2, R3, R4, and R5, for apple trees arranged in a $k \times k = 5 \times 5$ Latin square design and four soil treatments, 1, 2, 3, and 4, split blocked over the five rootstocks and over the five rows, and arranged in a randomized complete block design with the columns of the Latin square being the blocks. The layout of the experiment is shown below:

	Column				
	1	2	3	4	5
Soil treatment	4 2 3 1	2 3 1 4	4 3 1 2	4 2 1 3	1 2 3 4
Row 1	R3	R4	R5	R2	R1
Row 2	R2	R3	R1	R4	R5
Row 3	R4	R5	R2	R1	R3
Row 4	R1	R2	R3	R5	R4
Row 5	R5	R1	R4	R3	R2

The soil treatments were in a 2×2 factorial arrangement with no fumigation and fumigation and no composting and composting as the two factors. One response measured was height of apple trees grafted onto the five rootstocks. A partitioning of the degrees of freedom for the example, and in general for the above designed experiment is presented below:

Source of variation	Degrees of freedom ex.	Degrees of freedom general
Total	100	sk^2
Correction for the mean	1	1
Rows	4	$k-1$
Columns	4	$k-1$
Rootstocks	4	$k-1$
Error for rootstocks	12	$(k-1)(k-2)$
Soil treatments	3	$s-1$
Composting	1	$c-1$
Fumigation	1	$f-1$
Composting × fumigation	1	$(c-1)(f-1)$
Error for soil treatments = columns × soil treatments	12	$(k-1)(s-1)$
Rootstocks × soil treatments	12	$(k-1)(s-1)$
Rootstocks × composting	4	$(c-1)(k-1)$
Rootstocks × fumigation	4	$(f-1)(k-1)$
Rootstocks × composting × fumigation	4	$(c-1)(k-1)(f-1)$
Error for rootstocks × soil treatments	48	$(k-1)(k-1)(s-1)$

For the above analysis of variance table, soil treatments and rootstocks are considered to be fixed effects. SAS PROC GLM MODEL statement for the above ANOVA would be:

Response = row column rootstock soil soil ∗ column rootstock ∗ soil column ∗ rootstock ∗ soil

The "Error for rootstocks" is listed as the "Error" in the computer output. The term "Error for rootstocks × soil treatments" is obtained from "column ∗ rootstock ∗ soil" output. This is the error term for the interaction rootstock × soil interaction. The MODEL statement when composting C and fumigation F are added, is:

Response = row column rootstock C F C ∗ F soil ∗ column rootstock ∗ C rootstock ∗ F rootstock ∗ C ∗ F column ∗ rootstock ∗ soil

Note that either the ANOVA or the MODEL statement defines the parameters in a linear model used for obtaining an analysis of variance and the means.

4.3. BOTH SETS OF TREATMENTS IN SPLIT BLOCK ARRANGEMENTS

If the soil treatments had been designed to consider orders in the columns in orderings such as 4231, 2314, 1423, 3142, and 1234 to form a four order (row) and five column Youden design, the analysis of variance table would take the form:

Source of variation	Degrees of freedom
Total	$100 = sk^2$
Correction for the mean	1
Rows	$4 = k - 1$
Columns	$4 = k - 1$
Rootstocks	$4 = k - 1$
Error for rootstocks	$12 = (k - 1)(k - 2)$
Soil treatments	$3 = s - 1$
Composting	$1 = c - 1$
Fumigation	$1 = f - 1$
Composting × fumigation	$1 = (c - 1)(f - 1)$
Orders	$3 = o - 1$
Error for soil treatments	$9 = (k - 1)(s - 1) - (o - 1)$
Rootstocks × soil treatments	$12 = (s - 2)(k - 1)$
Rootstocks × composting	$4 = (k - 1)(c - 1)$
Rootstocks × fumigation	$4 = (k - 1)(f - 1)$
Rootstocks × composting × fumigation	$4 = (k - 1)(c - 1)(f - 1)$
Error for rootstocks × soil treatments	$48 = (s - 1)(k - 1)^2$

Note that a five column by four order experiment design is formed by adding any row of the 4 × 4 Latin square to the Latin square. This forms what is known as a Youden design, where the orders and soil treatments are in a balanced block arrangement.

4.4. SPLIT BLOCK SPLIT BLOCK OR STRIP STRIP BLOCK EXPERIMENT DESIGN

The experiment discussed in Section 4.2 was not the one used but had an additional set of treatments split blocked in the rows of the Latin square. One could call such an arrangement a *split block split block experiment design* or a *strip strip block experiment design*. The third set of treatments involved whether there had been an old apple tree occupying the area initially, or there was grass in the area. The arrangement of grass and old tree was systematic in every row of the Latin square; that is, the systematic arrangement was necessitated by the area available for the experiment. The plan of the experiment was:

	Column																			
	1				2				3				4				5			
Soil treatment	4	2	3	1	2	3	1	4	4	3	1	2	4	2	1	3	1	2	3	4
Row 1 Grass		R3				R4				R5				R2				R1		
Old tree																				
Row 2 Grass		R2				R3				R1				R4				R5		
Old tree																				
Row 3 Grass		R4				R5				R2				R1				R3		
Old tree																				
Row 4 Grass		R1				R2				R3				R5				R4		
Old tree																				
Row 5 Grass		R5				R1				R4				R3				R2		
Old tree																				

A linear model for this layout of an experiment is

$$Y_{fghij} = \mu + \rho_f + \pi_g + \alpha_h + \eta_{fgh} + \beta_i + \beta\pi_{gi} + \alpha\beta_{hi} + \omega_{ghi} + \gamma_j + \gamma\rho_{fj} + \gamma\pi_{gj} + \alpha\gamma_{hj} + \lambda_{fhj} + \beta\gamma_{ij} + \varphi_{fij} + \alpha\beta\gamma_{hij} + \varepsilon_{fghij},$$

where μ is a general mean effect,

ρ_f is the fth random row effect distributed with mean zero and variance σ_ρ^2,

π_g is the gth random column effect identically and independently distributed with mean zero and variance σ_π^2,

α_h is the effect of the hth rootstalk,

η_{fgh} is a random error effect identically and independently distributed (IID) with mean zero and variance σ_r^2,

β_i is the effect of the ith soil treatment,

$\beta\pi_{gi}$ is a random error effect IID with mean zero and variance σ_s^2,

$\alpha\beta_{hi}$ is an interaction effect of the hth rootstalk and the ith soil treatment,

ω_{ghi} is a random error effect IID with mean zero and variance σ_ϖ^2,

γ_j is the effect of the jth previous treatment,

$\gamma\rho_{fj}$ is a random error effect IID with mean zero and variance $\sigma_{\gamma\rho}^2$,

$\gamma\pi_{gj}$ is a random error effect of the jth previous treatment with the gth column IID with mean zero and variance $\sigma_{\gamma\pi}^2$,

$\alpha\gamma_{hj}$ is an interaction effect of rootstalk h with previous treatment j,

λ_{fhj} is a random error effect IID with mean zero and variance σ_λ^2,

$\beta\gamma_{ij}$ is an interaction effect of soil treatment i and previous treatment j,

φ_{fij} is a random error effect IID with mean zero and variance σ_φ^2,

$\alpha\beta\gamma_{hij}$ is a three factor interaction effect of the hth rootstalk with the ith soil treatment and jth previous treatment,

ε_{fghij} is a random error effect IID with mean zero and variance σ_ε^2.

The random effect terms are assumed to be mutually independent.

Given the above linear model, a partitioning of the degrees of freedom in an analysis of variance table is shown below:

Source of variation	Degrees of freedom
Total	$200 = spk^2$
Correction for the mean	1
Rows	$4 = k - 1$
Columns	$4 = k - 1$
Rootstocks $= R$	$4 = k - 1$
Error for rootstocks	$12 = (k-1)(k-2)$
Soil treatments $= S$	$3 = s - 1$
Composting	$1 = c - 1$
Fumigation	$1 = f - 1$
Composting × fumigation	$1 = (c-1)(f-1)$
Error for soil treatments = columns × soil treatments	$12 = (k-1)(s-1)$
Rootstocks × soil treatments	$12 = (k-1)(s-1)$
Rootstocks × composting	$4 = (k-1)(c-1)$
Rootstocks × fumigation	$4 = (k-1)(f-1)$
Rootstocks × composting × fumigation	$4 = (k-1)(c-1)(f-1)$
Error for rootstocks × soil treatments	$48 = (s-1)(k-1)^2$
Grass versus old tree $= P$	$1 = p - 1$
Row × P = error for P	$4 = (k-1)(p-1)$
Columns × P	$4 = (k-1)(p-1)$
$R \times P$	$4 = (k-1)(p-1)$
Error $R \times P$	$12 = (k-1)(k-2)(p-1)$
$S \times P$	$3 = (s-1)(p-1)$
Error $S \times P$	$12 = (s-1)(k-1)(p-1)$
$S \times R \times P$	$12 = (s-1)(k-1)(p-1)$
Error for $S \times R \times P$	$48 = (s-1)(p-1)(k-1)^2$

When presented with the above key-out of the degrees of freedom, the experimenter wondered about the rationale for the above analysis. In order to comprehend a complete breakdown of the degrees of freedom in an analysis of variance, it is useful to break down the analysis for separate parts of the experiment. One way to do this is to consider the responses from the grass alone part of the experiment (See Chapter 2). This plus the other half of the experiment on the responses from the old tree part of the experiment are given below. For grass alone, the experiment is designed as a split block with one set of treatment in a randomized complete block design and the other set in a Latin square design. The same design holds for the old tree part of the experiment. These partitionings of the degrees of freedom for an analysis of variance were explained to the analyst as follows:

Source of variation	Degrees of freedom	
	Grass	Old tree
Total	100	100
Correction for the mean	1	1
Rows	4	4
Columns	4	4
Rootstocks	4	4
Error for rootstocks	12	12
Soil treatments	3	3
Error for soil treatments = columns × soil treatments	12	12
Rootstocks × soil treatments	12	12
Error for rootstocks × soil treatments	48	48

The Error for rootstock × soil treatment may be computed as the rootstock × soil treatment × column interaction. When the above analyses are understood, one is in a position to prepare a combined analysis. Let the symbol P denote the previous land use factor-grass and old tree. From combining the above analyses, the partitioning of degrees of freedom for the combined experiment is outlined below:

Source of variation	Degrees of freedom
Total	200
Correction for the mean within P	2
Correction for mean	1
P	1
Rows within P	8
Columns within P	8
Columns	4
Columns × P	4
Rootstocks within P	8
Rootstocks = R	4
$R \times P$	4
Error R within P	24
Error for rootstocks	12
Error for $R \times P$	12
Soil treatments within P	6
Soil treatments = S	3
$S \times P$	3
Error for S within P	24
Error for soil treatments	12
Error for $S \times P$	12
$R \times S$ within P	24
$R \times S$	12
$R \times S \times P$	12
Error for $R \times S$ within P	96
Error for $R \times S = R \times S \times$ column	48
Error for $R \times S \times P = R \times S \times P \times$ col.	48

There are six different error terms in the above analysis of variance table. There would have been seven if grass and old tree had been randomized. There were three different error terms each, for grass and old tree responses alone. Even though this was a logical way to lay out the experiment, such arrangements add to the complexity of statistical analysis and decisions on appropriate error terms, for the various effects. The benefit is the additional amount of information that is available.

Example 4.1—A numerical example using artificial data for the above described experiment is given in Table 4.1 for grass data only (see appendix 4.1). The symbol P stands for previous treatment, the symbol R for rootstock, and the symbol S for soil treatment.

For the data from both the grass and old tree parts of the experiment, the SAS code for obtaining an analysis of variance and the treatment means is given in Appendix 4.1. The analysis of variance tables for the two SAS PROC GLM runs are given in Tables 4.2 and 4.3.

The error sum of squares for rootstocks, Error R, is obtained by subtracting the row sum of squares, 309.430, with 4 degrees of freedom, from the column*R sum of squares, 467.920, with 16 degrees of freedom, and is equal to 158.490 with 12 degrees of freedom. Thus, the Error R mean square is equal to 158.490 / 12 = 13.207.

Table 4.1. Data for a Split Block Split Block Designed Experiment with Five Rootstocks, R, in a 5 × 5 Latin Square Design, Four Soil Treatments, S, in a Randomized Complete Block Design with Columns as Replicates, and Two Previous Managements, P is for grass data only.

S	S	S	S	S	
4 2 3 1	2 3 1 4	3 1 2 4	2 1 3 4	1 2 3 4	P
R3	R4	R5	R2	R1	
13 8 11 11	10 8 10 9	9 9 10 7	9 12 9 10	12 17 8 9	Grass
R2	R3	R1	R4	R5	
9 7 8 5	9 8 6 3	7 9 5 8	7 5 9 4	8 3 8 6	Grass
R4	R5	R2	R1	R3	
9 8 8 6	8 9 2 8	8 5 8 5	7 7 6 8	8 5 8 8	Grass
R1	R2	R3	R5	R4	
0 1 4 0	0 0 0 0	9 8 7 9	5 5 5 6	9 5 8 8	Grass
R5	R1	R4	R3	R2	
8 8 9 7	8 7 8 9	8 7 8 8	8 5 7 9	8 8 5 9	Grass

Table 4.2. Type I (Also Type III) ANOVA and Associated F-Values for the First Model.

Source of variation	Degrees of freedom	Sum of squares	Mean square	F-value
Total, corrected	199	1710.155		
P	1	28.125	28.125	1.23
Column	4	34.330	8.582	
Column × P	4	91.450	22.862	
R	4	31.030	7.758	0.59
P × R	4	48.950	12.238	0.56
Column × R	16	467.920	29.245	
Column × P × R	16	350.100	21.881	
S	3	3.775	1.258	0.20
P × S	3	3.295	1.098	0.28
Column × S	12	74.550	6.212	
Column × P × S	12	47.030	3.919	
R × S	12	36.650	3.054	0.74
Column × R × S	48	197.400	4.112	
P × R × S	12	26.330	2.194	0.39
Column × P × R × S	48	269.220	5.609	

To make some of the desired F-tests in an analysis of variance, the following statements may be inserted after the SAS code MODEL statement that is presented in Appendix 4.1:

$$\text{Test H} = S^*R^*P \ E = S^*R^*P^*\text{column};$$
$$\text{Test H} = S^*P \ E = S^*P^*\text{column};$$
$$\text{Test H} = R^*P \ E = \text{row}^*R^*P;$$
$$\text{Test H} = S^*R \ E = R^*S^*\text{column};$$
$$\text{Test H} = S \ E = \text{column}^*S;$$

The F-test for rootstock effects, $F = 7.758 / 13.207$, will have to be made manually.

Table 4.3. Type I and Type III ANOVA and Associated F-Values for the Second Model.

Source of variation	Degrees of freedom	Sum of squares	Mean square	F-value
Total, corrected	199	1710.155		
Row	4	309.430	77.358	
R	4	31.030	7.758	
P	1	28.125	28.125	0.86
S	3	3.775	1.258	
R × S	12	36.650	3.054	0.23
Row × P	4	130.250	32.562	
P × R	4	48.950	12.238	0.78
Row × P × R	32	504.120	15.754	
P × S	3	3.295	1.098	0.15
Row × P × S	24	171.280	7.137	
P × R × S	12	26.330	2.194	0.51
Row × P × R × S	96	419.920	4.343	

The means for rootstock, previous treatment, and soil treatment means, and means for combinations of these treatments are given in the abbreviated SAS output presented in Appendix 4.1.

4.5. ONE SET OF TREATMENTS IN AN INCOMPLETE BLOCK DESIGN AND THE SECOND SET IN A RANDOMIZED COMPLETE BLOCK DESIGN

In some instances, it is desirable to split block an experiment designed as an incomplete block experiment design. Suppose there are 12 genotypes, factor A, treatments designed as an incomplete block design in $s = 4$ incomplete blocks of size $k = 3$, and that there are $r = 4$ complete blocks or replicates of the $a = sk = 12$ genotypes. And suppose that a second set of $b = 3$ herbicides, factor B treatments, are to be used in a split block arrangement. These would be across the 12 genotype experimental units in the four incomplete blocks of each complete block. A schematic layout of such an experiment is given below. The genotype numbers listed inside the parentheses indicate the incomplete blocks and the genotypes appearing in each incomplete block of the incomplete block experiment design.

Replicate 1

(1 2 3)	(4 5 6)	(7 8 9)	(10 11 12)	
				B1
				B2
				B3

Replicate 2

(1 5 9)	(4 8 12)	(7 11 3)	(10 2 6)	
				B1
				B2
				B3

Replicate 3

(1 8 6)	(4 11 9)	(7 2 12)	(10 5 3)	
				B1
				B2
				B3

Replicate 4

(1 4 7)	(5 8 11)	(3 6 9)	(2 10 12)	
				B1
				B2
				B3

A partitioning of the degrees of freedom in an analysis of variance is shown in the following table:

Source of variation	Degrees of freedom
Total	$abr = rbsk = 144$
Correction for the mean	1
Replicate	$r - 1 = 3$
Genotype = A	$a - 1 = 11$
Incomplete blocks within replicate	$r(s - 1) = 12$
Intrablock error or residual	$(r - 1)(a - 1) - r(k - 1) = 33 - 8 = 25$
Herbicide = B	$b - 1 = 2$
Error $B = B \times$ replicate	$(b - 1)(r - 1) = 6$
$A \times B$	$(a - 1)(b - 1) = 22$
Error $A \times B = A \times B \times$ replicate	$(a - 1)(b - 1)(r - 1) = 66$

Incomplete block information would ordinarily be recovered in such an experiment as this. To do this, use may be made of SAS PROC MIXED or other comparable software. Standard errors of a difference between two genotype means adjusted for recovery of interblock information would be used for comparisons of genotype effects. The herbicide effect would be tested using Error B, and the genotype × herbicide interaction effect would be tested making use of the Error $A \times B$ mean square. A SAS PROC GLM MODEL statement for the above ANOVA table would be

Response = replicate A block(replicate) B B*replicate A*B A*B*replicate

The "Error" term in the SAS PROC GLM output for the above model would be the Intrablock error or residual for factor A.

4.6. AN EXPERIMENT DESIGN SPLIT BLOCKED ACROSS THE ENTIRE EXPERIMENT

Suppose an experimenter is planning to conduct an experiment in an area where there had been $b = 4$ previous crops, factor B. It is desired to determine the effect on $a = 12$ genotypes, factor A, of the previous cropping use on genotype response. Suppose the schematic arrangement of the incomplete block experiment design was that of the previous section but was laid out as follows:

Replicate 1			Replicate 2			Replicate 3			Replicate 4			B
1	2	3	1	5	9	1	8	3	1	4	7	B1
4	5	6	4	8	12	4	11	6	5	8	11	B2
7	8	9	7	11	3	7	2	9	3	6	9	B3
10	11	12	10	2	6	10	5	12	2	10	12	B4

This would not be a desirable experiment design. The first one of the incomplete blocks falls on one previous cropping area and likewise for the three remaining rows of incomplete blocks. This leads to confounding between the previous cropping systems and the incomplete blocks. Also, there is no orthogonality between genotypes and previous crops. In running an analysis on the 48 genotype by replicate responses over previous crops, the main effect of the previous crop would be taken care of by the incomplete blocks, but the previous crop by genotype interaction would not.

Suppose, however, the following schematic layout of an experiment was used for $a = 4$ levels of factor A arranged in a randomized complete block design with $r = 3$ replicates. Furthermore, suppose that $b = 4$ levels of factor B, were split blocked over the entire experiment as follows:

Replicate 1				Replicate 2				Replicate 3				B
1	2	3	4	1	2	3	4	1	2	3	4	B1
1	2	3	4	1	2	3	4	1	2	3	4	B2
1	2	3	4	1	2	3	4	1	2	3	4	B3
1	2	3	4	1	2	3	4	1	2	3	4	B4

A partitioning of the degrees of freedom in an analysis of variance for this layout would be:

Source of variation	Degrees of freedom
Total	48
Correction for the mean	1
Factor B	3
Replicate	2
Replicate × factor B	6
Factor A	3
Replicate × Factor A = Error A	6
Factor A × factor B	9
Replicate × A × B = Error A × B	18

Since the factor B treatments are unreplicated, there is no error term for comparing these effects. The error terms for factor A and its interaction with factor B are the usual ones.

4.7. CONFOUNDING IN A FACTORIAL TREATMENT DESIGN AND IN A SPLIT BLOCK EXPERIMENT DESIGN

Instead of having the main effects of a factorial treatment design as the whole plots of a split block experiment design, one may use interactions as the whole plots. This is illustrated using the following three examples. As a first example, suppose one desires to conduct a 2^3 factorial experiment of factors A, B, and C, with the interactions AB and AC as the whole plot treatments. It may be that the experimenter has little interest in the interactions AB and AC or may believe that they are negligible or nonexistent. A schematic layout of one of r replicates of this arrangement would be:

	AC_0		AC_1	
AB_0	000	111	101	110
AB_1	101	010	011	100

AB would be in a randomized complete block design as would AC. The two combinations within each intersection of the levels of the interactions would be randomly allocated to the two experimental units. For r replications of this plan, an analysis of variance key-out of the degrees of freedom would be as follows:

Source of variation	Degrees of freedom
Total	$8r$
Correction for mean	1
Replicates $= R$	$r - 1$
$A \times B$	1
$A \times B \times R$	$r - 1$
$A \times C$	1
$A \times C \times R$	$r - 1$
A	1
B	1
C	1
$B \times C$	1
$A \times B \times C$	1
Error	$5(r - 1)$

The $A \times B \times R$ mean square is the error term for testing for an $A \times B$ interaction. The $A \times C \times R$ mean square would be used as the error term for the $A \times C$ interaction. The Error mean square would be the appropriate one for comparing the remaining factorial effects.

As a second example, consider a $2 \times 2 \times 3$ factorial treatment design for factors A, B, and C, respectively, and consider that interactions $A \times B$ and $A \times C$ are not as important as the other factorial effects. The following is a schematic plan for one replicate:

	AC_0		AC_1		AC_2	
AB_0	000	112	001	110	002	111
AB_1	010	102	011	100	012	101

A key-out of the degrees of freedom in an analysis of variance table for r replicates is presented below:

Source of variation	Degrees of freedom
Total	$12r$
Correction for mean	1
Replicate	$r - 1$
$A \times B$	1
$A \times B \times R$ = error AB	$r - 1$
$A \times C$	2
$A \times C \times R$ = error AC	$2(r - 1)$
A	1
B	1
C	2
$B \times C$	2
$A \times B \times C$	2
Error	$8(r - 1)$

The "Error" mean square is the appropriate error term for the last five effects in the above table.

As a third illustration of using interaction effects as whole plot treatments in a split block designed experiment, consider the $2 \times 3 \times 3$ factorial treatment design for the factors A, B, and C, respectively. Suppose that the experimenter believes that the $B \times C$ interaction is not as important as the other effects in this factorial arrangement. The levels of the geometrical components of the $B \times C$ interaction, that is, BC and BC^2 (See e.g., Federer, 1955, Chapter VII; Kempthorne, 1952; for a description and construction of geometrical components of an interaction), would form the two whole plot treatments. A schematic arrangement for this plan for one replicate would be:

	BC_0	BC_1	BC_2
BC_0^2	000	022	011
	100	122	111
BC_1^2	021	010	002
	121	110	102
BC_2^2	012	001	020
	112	101	120

A key-out of the degrees of freedom in an analysis of variance for r replicates of a split block experiment design would be:

Source of variation	Degrees of freedom
Total	$18r$
Correction for mean	1
Replicate = R	$r-1$
BC	2
$BC \times R$ = error BC	$2(r-1)$
BC^2	2
$BC^2 \times R$ = error BC^2	$2(r-1)$
A	1
B	2
$A \times B$	2
C	2
$A \times C$	2
$A \times B \times C$	4
Error	$13(r-1)$

The "Error" mean square is the appropriate error mean square for the last six factorial effects in the above table. Note that the levels of the geometric components of the three factor interaction $A \times B \times C$ could have been used as the whole plot treatments.

The usual randomization procedures for a split block experiment design apply equally well for arrangements such as those discussed above.

4.8. SPLIT BLOCK EXPERIMENT DESIGN WITH A CONTROL

Mejza (1998) has presented a class of split block experiment designs wherein a control treatment represents one of the treatments for either or both of the two factors involved. That is, factor A may include a control treatment in every split block experimental unit, sbeu, but the sbeu treatments change from block to block. Likewise, a control treatment may be included in every sbeu for factor B. The following example with data is given by Mejza (1998), as an illustration of a member of this class of experiment designs where $A1$ is the control treatment:

	Block 1			Block 2			Block 3			Block 4	
	$A1$	$A2$		$A3$	$A1$		$A4$	$A1$		$A1$	$A5$
$B2$	5.37	21.04	$B1$	19.26	1.92	$B1$	18.22	1.15	$B1$	2.10	9.86
$B1$	1.01	14.87	$B2$	26.46	9.97	$B2$	22.23	6.01	$B2$	14.00	27.78

	Block 5			Block 6			Block 7			Block 8	
	$A2$	$A1$		$A3$	$A1$		$A1$	$A4$		$A5$	$A1$
$B2$	27.29	5.97	$B1$	16.48	2.20	$B1$	1.18	17.98	$B2$	24.50	6.14
$B1$	24.30	1.29	$B2$	26.22	14.66	$B2$	8.09	25.15	$B1$	14.24	1.10

The control $A1$ is replicated eight times whereas $A2$ to $A5$ are only replicated twice. $B1$ and $B2$ treatments are in a standard split block arrangement of eight blocks. A Type I analysis of variance for this data set is given below:

Source of variation	Degrees of freedom	Sum of squares	Mean square
Total	32	8,216.72	
Correction for mean	1	5,461.17	
Block	7	72.00	10.29
A	4	2,042.27	510.57
$A \times$ block	4	47.80	11.95
B	1	478.33	478.33
$B \times$ block	7	95.51	13.64
$A \times B$	4	17.43	4.36
$A \times B \times$ block	4	2.20	0.55

A Type III analysis of variance is the same as given above. The $A \times$ block mean square is used as the error term for factor A, the $B \times$ block mean square is the error for factor B, and the $A \times B \times$ block mean square is the error term for the $A \times B$ interaction. A SAS MODEL statement to obtain the above analysis is:

Yield = block A A*block B B*block A*B A*B*block;

Using only blocks 1 to 4 or 5 to 8 would not result in a solution for all of the effects. However, adding more levels of factor A to each of the blocks in a fashion similar to the following does result in solutions for effects:

	Block 1				Block 2		
	A1	A2	A3		A3	A1	A4
B2	5.37	21.04	27.29	B1	19.26	1.92	5.97
B1	1.01	14.87	24.30	B2	26.46	9.97	1.29
	Block 3				Block 4		
	A4	A1	A5		A1	A5	A2
B1	18.22	1.15	2.10	B1	2.10	9.86	24.50
B2	22.23	6.01	14.00	B2	14.00	27.78	14.24

A Type III analysis of variance for the above data set is:

Source of variation	Degrees of freedom	Sum of squares	Mean square
Block	3	164.56	54.85
A	4	1,159.49	289.87
Error A	4	268.13	67.03
B	1	147.39	147.39
$B \times$ block, Error B	3	3.10	1.03
$A \times B$	4	177.28	44.32
Error AB	4	114.92	28.73

The same error terms and SAS model statements apply here as for the previous data set.

This design has some similarities to augmented experiment designs (See Federer, 1993, 2002, and related references therein). Usually the "new" treatments are only included once in an augmented experiment design. In Mejza's (1998) design, $A2$ to $A5$ were included twice while the control, $A1$, was replicated eight times. One could use the above ideas to construct an *augmented split block experiment design* as described in Chapter 8. For example, suppose an experimenter used four standard or control genotypes and 64 new genotypes as the A factor and two levels of fertilizers with two types of soil preparation as the B factor. It was desired to test or screen the new treatments on the four B treatments. Further, suppose that the design for the B factor was a randomized complete block experiment design and that an incomplete block design with four incomplete blocks of size two was used for the A control treatments. A plan before randomization for this would be:

	Block 1		Block 2		Block 3		Block 4	
	A1	A2	A2	A3	A3	A4	A4	A1
B1								
B2								
B3								
B4								

In each of the above incomplete blocks, the incomplete block size would be expanded and sixteen of the 64 new treatments would be included. A set of sixteen new and two controls would be randomly allotted to the eighteen experimental units for factor A experimental units, in each of the four incomplete blocks. The B treatments would run across these eighteen factor A treatments in each incomplete block. The experiment design for factor B would be a randomized complete block. If more new treatments were to be tested, the number of blocks could be increased and/or more new treatments could be included with the factor A experimental units in a block.

A partitioning of the degrees of freedom in an analysis of variance table for the above designed experiment would be as follows:

Source of variation	Degrees of freedom
Total	288
Correction for mean	1
Block	3
A treatments	67
Control or standard	3
Control versus new	1
New	63
Error A	1
B	3

(*Continued*)

(*Continued*)

Source of variation	Degrees of freedom
B × block, Error B	9
A × B	201
B × control	9
B × control versus new	3
B × new	189
Error AB	3

Though the above design is connected, there are insufficient degrees of freedom associated with the error terms. Hence, additional blocks and/or controls will be required in order to obtain the desired degrees of freedom for the various error terms.

4.9. COMMENTS

From the above examples on the variations of split block designs, the analyst should be able to determine an appropriate key-out of the degrees of freedom for other variations. There are many more variations that have been and will be used by experimenters. For the rootstock by soil treatment by previous treatment example, described before, the design used is a perfectly reasonable one as far as the experimenter was concerned. Federer (1984) stated an axiom of experimentation as

Design for the experiment, do not experiment for the design.

The experiment should be conducted and designed to meet the goals of the experimenter, and not to obtain a simple and easy statistical analysis or to fit into some design with which a statistician or an experimenter is familiar.

4.10. PROBLEMS

Problem 4.1. For the numerical example in Section 4.8, obtain a stem and leaf diagram of the residuals. Does this suggest another model?

Problem 4.2. Obtain the residuals for the data of Example 4.1. Does a study of the residuals suggest another model?

Problem 4.3. Compare the results obtained for the problems in Chapter 2 with those obtained in Problem 4.2.

Problem 4.4. Artificial data were obtained for the first two replicates of the experiment design described in Section 4.5. The data are presented in the following table:

	Replicate 1											
(1	2	3)	(4	5	6)	(7	8	9)	(10	11	12)	
28	34	24	20	22	37	20	29	29	23	25	23	B1
25	23	33	33	26	27	33	34	26	27	28	35	B2
35	29	32	26	40	20	20	32	29	22	35	24	B3
	Replicate 2											
(1	5	9)	(4	8	12)	(7	11	3)	(10	2	6)	
24	36	32	26	21	39	36	32	26	21	39	40	B1
25	20	21	26	36	28	37	29	30	30	27	23	B2
34	35	31	21	34	37	40	37	40	28	20	24	B3

(i) Write a computer code for obtaining an analysis of the above data.

(ii) Obtain the factor *A* means both with and without recovery of interblock information.

4.11 REFERENCES

Federer, W. T. (1955). *Experimental Design: Theory and Application*, The Macmillan Company, New York, pp. 1–544.

Federer, W. T. (1984). Principles of statistical design with special reference to experiment and treatment design. In *Statistics: An Appraisal—Proceedings of the 50th Anniversary Conference, Iowa State Statistics Laboratory* (Editors: H. A. David and H. T. David), Iowa State Press Ames, Iowa, pp. 77–104.

Federer, W. T. (1993). *Statistical Design and Analysis for Intercropping Experiments: Volume I Two Crops*, Springer-Verlag, New York, Heidelberg, Berlin, Chapter 10, pp. 1–298.

Federer, W. T. (2002). Construction and analysis of an augmented lattice square design. Biometric J 44(2):251–257.

Kempthorne, O. (1952). *The Design and Analysis of Experiments*, John Wiley & Sons, Inc., New York, pp. 1–631.

Mejza, Iwona (1998). Characterization of certain split-block designs with a control. Biometric J 40(5):627–639.

APPENDIX 4.1. EXAMPLE 4.1 CODE

The data format and computer code for Example 4.1 are given below. The complete data set is given on the book's FTP site.

```
data example;
input row P column R S height;
datalines;
1       1       1       3       4       13
1       1       1       3       2       8
```

```
1          1          1          3          3          11
1          1          1          3          1          11
...

5          2          5          2          2          6
5          2          5          2          3          8
5          2          5          2          4          9
;
RUN;
PROC GLM DATA = example;
CLASS row column P R S;
MODEL height = P column column*P R P*R column*R column*R*P S
  P*S column*S column*S*P
R*S R*S*column R*S*P R*S*P*column;
LSMEANS R S P R*S R*P P*S;
RUN;
PROC GLM data = example;
CLASS row column P R S;
MODEL height = row R P S S*R row*P R*P row*R*P S*P
S*P*row S*R*P R*S*P*row;
LSMEANS R S P;
RUN;
```

An abbreviated form of the output of the above code and data set is given below:

```
Dependent Variable: height
                          Sum of
Source              DF    Squares       Mean Square   F Value  Pr > F
Model              199   1710.155000     8.593744       .       .
Error                0      0.000000                    .
Corrected Total    199   1710.155000

              R-Square   Coeff Var   Root MSE   height Mean
              1.000000      .            .        6.815000

Source              DF    Type I SS    Mean Square   F Value  Pr > F
P                    1    28.1250000   28.1250000       .       .
column               4    34.3300000    8.5825000       .       .
column*P             4    91.4500000   22.8625000       .       .
R                    4    31.0300000    7.7575000       .       .
P*R                  4    48.9500000   12.2375000       .       .
column*R            16   467.9200000   29.2450000       .       .
column*P*R          16   350.1000000   21.8812500       .       .
S                    3     3.7750000    1.2583333       .       .
P*S                  3     3.2950000    1.0983333       .       .
column*S            12    74.5500000    6.2125000       .       .
```

APPENDIX

```
column*P*S      12    47.0300000    3.9191667    .    .
R*S             12    36.6500000    3.0541667    .    .
column*R*S      48   197.4000000    4.1125000    .    .
P*R*S           12    26.3300000    2.1941667    .    .
column*P*R*S    48   269.2200000    5.6087500    .    .
```

Least Squares Means

height

R	LSMEAN
1	7.20000000
2	6.07500000
3	7.00000000
4	6.77500000
5	7.02500000

height

S	LSMEAN
1	6.58000000
2	6.92000000
3	6.86000000
4	6.90000000

height

P	LSMEAN
1	7.19000000
2	6.44000000

height

R	S	LSMEAN
1	1	7.60000000
1	2	8.10000000
1	3	6.50000000
1	4	6.60000000
2	1	5.20000000
2	2	6.20000000
2	3	6.20000000
2	4	6.70000000
3	1	6.80000000
3	2	7.10000000
3	3	7.10000000
3	4	7.00000000
4	1	6.40000000
4	2	6.80000000

```
4   3   6.90000000
4   4   7.00000000
5   1   6.90000000
5   2   6.40000000
5   3   7.60000000
5   4   7.20000000
```

Least Squares Means

Height

```
P   R   LSMEAN

1   1   7.00000000
1   2   6.25000000
1   3   8.00000000
1   4   7.70000000
1   5   7.00000000
2   1   7.40000000
2   2   5.90000000
2   3   6.00000000
2   4   5.85000000
2   5   7.05000000
```

height

```
P   S   LSMEAN

1   1   6.88000000
1   2   7.16000000
1   3   7.44000000
1   4   7.28000000
2   1   6.28000000
2   2   6.68000000
2   3   6.28000000
2   4   6.52000000
```

Dependent Variable: height

Source	DF	Sum of Squares	Mean Square	F Value	Pr > F
Model	199	1710.155000	8.593744	.	.
Error	0	0.000000	.		
Corrected Total	199	1710.155000			

R-Square	Coeff Var	Root MSE	height Mean
1.000000	.	.	6.815000

Source	DF	Type I SS	Mean Square	F Value	Pr > F
row	4	309.4300000	77.3575000	.	.
R	4	31.0300000	7.7575000	.	.
P	1	28.1250000	28.1250000	.	.

S	3	3.7750000	1.2583333	.	.
R*S	12	36.6500000	3.0541667	.	.
row*P	4	130.2500000	32.5625000	.	.
P*R	4	48.9500000	12.2375000	.	.
row*P*R	32	504.1200000	15.7537500	.	.
P*S	3	3.2950000	1.0983333	.	.
row*P*S	24	171.2800000	7.1366667	.	.
P*R*S	12	26.3300000	2.1941667	.	.
row*P*R*S	96	416.9200000	4.3429167	.	.

Least Squares Means

height

R	LSMEAN
1	7.20000000
2	6.07500000
3	7.00000000
4	6.77500000
5	7.02500000

height

S	LSMEAN
1	6.58000000
2	6.92000000
3	6.86000000
4	6.90000000

height

P	LSMEAN
1	7.19000000
2	6.44000000

CHAPTER 5

Combinations of SPEDs and SBEDs

5.1 INTRODUCTION

There are situations where it is necessary and/or desirable to use combinations of split plot and split block experiment designs (SBEDs). Such a designed experiment presents challenges to the analyst in producing an appropriate statistical analysis of the data from the experiment. Some of these challenges are described in this and the next chapter. The first such combination of these two designs is described in Section 5.2. Here, a standard split block experiment design is used for factors A and B and then the factor C is in a split plot arrangement to these factors. In Section 5.3, factor A constitutes the whole plot treatment; factors B and C are the split plot treatments but are in a split block arrangement within each whole plot treatment. In Section 5.4, factors A and B are in a standard split plot arrangement and factor C is in a split block arrangement to the split and whole plots of factors A and B. There are six different error terms in the statistical analysis of an experiment designed in this fashion. In Section 5.5, 55 rows by 40 vines of a vineyard were used to run a five factor experiment involving types of root, nitrogen fertilization, method of management, method of pruning, and method of thinning. This complexly designed experiment produced a challenge in finding an appropriate statistical analysis of the data. Some rules are set forth in Section 5.6 as an aid in finding appropriate statistical analyses for complexly designed experiments. Some additional comments on the topic of this chapter are presented in Section 5.7.

5.2 FACTORS A AND B IN A SPLIT BLOCK EXPERIMENT DESIGN AND FACTOR C IN A SPLIT PLOT ARRANGEMENT TO FACTORS A AND B

Let factors A and B be in a split block arrangement in each of the r replicates of a randomized complete block design. This was called a standard split block experiment

---

Variations on Split Plot and Split Block Experiment Designs, by Walter T. Federer and Freedom King
Copyright © 2007 John Wiley & Sons, Inc.

FACTORS A AND B IN A SPLIT BLOCK EXPERIMENT DESIGN

design in Chapter 2. Then divide each combination of factors A and B into c split plot experimental units. The c levels of factor C are randomly allotted to the c speus within each of the ab combinations of factors A and B in each of the replicates. There are r randomizations of the levels of factor A and also r randomizations of the levels of the B factor. There are rab randomizations of the factor C treatments. The treatment design is usually a three-factor factorial. A linear model for this experiment design is

$$Y_{hij} = \mu + \rho_g + \alpha_h + \delta_{gh} + \beta_i + \pi_{gi} + \alpha\beta_{hi} + \eta_{ghi} + \gamma_j + \alpha\gamma_{hj} + \beta\gamma_{ij} + \alpha\beta\gamma_{hij} + \varepsilon_{ghij}$$

where $g = 1, 2, \ldots, r$, $h = 1, 2, \ldots, a$, $i = 1, 2, \ldots, b$, and $j = 1, 2, \ldots, c$,

μ = general overall mean effect,

ρ_g = gth replicate or block effect identically and independently distributed with mean zero and variance σ_ρ^2,

α_h = effect of hth level of factor A,

δ_{gh} = ghth random error effect identically and independently distributed, IID, with mean zero and variance σ_δ^2,

β_i = effect of ith level of factor B,

π_{gi} = gith random error effect IID with mean zero and variance σ_π^2,

$\alpha\beta_{hi}$ = interaction effect hith combination of factors A and B,

η_{ghi} = ghith randon error effect IID with zero mean and variance σ_η^2,

γ_j = effect of jth level of factor C,

$\alpha\gamma_{hj}$ = interaction effect of hjth combination of factors A and C,

$\beta\gamma_{ij}$ = interaction effect of ijth combination of factors B and C,

$\alpha\beta\gamma_{hij}$ = interaction effect of hijth combination of factors A, B, and C, and

ε_{ghij} = ghijth random error effect IID with zero mean and variance σ_ε^2.

The random effects $\rho_g, \delta_{gh}, \pi_{gi}, \eta_{ghi},$ and ε_{ghij} are assumed to be mutually independent.

Let $r = 4$ replicates, $a = 4$ levels of factor A, $b = 3$ levels of factor B, and $c = 4$ levels of factor C. Then a systematic layout of one replicate of this design would be as follows:

	A1	A2	A3	A4
B1	C1 C2 C3 C4	C1 C2 C3 C4	C1 C2 C3 C4	C1 C2 C3 C4
B2	C1 C2 C3 C4	C1 C2 C3 C4	C1 C2 C3 C4	C1 C2 C3 C4
B3	C1 C2 C3 C4	C1 C2 C3 C4	C1 C2 C3 C4	C1 C2 C3 C4

The levels of factor A would be randomized within each of the r blocks. The levels of factor B would be independently randomized within each of the r blocks. The levels of factor C would be randomized within each combination of levels of factors A and B and within each of the r blocks.

An analysis of variance partitioning of the degrees of freedom for this experiment design for the general case and for the example is

Source of variation	Degrees of freedom	
	General	Example
Total	$abcr$	$192 = 4(4)(3)(4)$
Correction for mean	1	1
Replicate or complete block $= R$	$r - 1$	3
Factor $A = A$	$a - 1$	3
$A \times R =$ error A	$(a-1)(r-1)$	9
Factor $B = B$	$b - 1$	2
$B \times R =$ error B	$(b-1)(r-1)$	6
$A \times B$	$(a-1)(b-1)$	6
$A \times B \times R =$ error AB	$(a-1)(b-1)(r-1)$	18
Factor $C = C$	$c - 1$	3
$A \times C$	$(a-1)(c-1)$	9
$B \times C$	$(b-1)(c-1)$	6
$A \times B \times C$	$(a-1)(b-1)(c-1)$	18
$C \times R$ within A and $B =$ error C	$ab(c-1)(r-1)$	108

There are four error terms in this analysis. The different experimental units and randomizations for factors A, B, and C differ, hence, giving rise to the four error terms-error A, error B, error AB, and error C. The experimental units for the interaction of factors A and B are different from those for the levels of factors A and B, giving rise to error AB. The experimental units for the levels of factor C are in a split plot arrangement for a combination of levels of factors A and B, giving rise to error C.

A further variant of this type of experiment design is discussed by Yates (1933). This classic and fundamental paper contains much more than a discussion of the design and the example presented below. For example, he gives the following definition of orthogonality. *Orthogonality is that property of the design which ensures that the different classes of effects to which the experimental material is subject shall be capable of direct and separate estimation without any entanglement.* This definition is from the estimation of effects viewpoint. Federer (1984, 1991) gives a combinatorial definition of orthogonality as follows: *If the proportions of levels, L1:L2:L3: . . ., of a factor remain the same for all levels of a second factor, the factors are said to form an orthogonal arrangement. For n factors, the arrangement must be pairwise orthogonal.* The combinatorial definition was found to be intelligible to first year college students, and the definition of estimation was not as they had no knowledge or grasp of the estimability concept. Also, the combinatorial definition runs parallel to that of a balanced incomplete block experiment design.

One of the designs discussed by Yates (1933) is to use the rows and columns as the whole plot experimental units (wpeus) rather than splitting the rows and/or columns into wpeus as was done for Example 2.2 and Problem 2.3. An example of a schematically arranged design with rows as whole plots and with the split plot treatments in a 6 × 6 Latin square arrangement is given below for six rows forming three pairs, R1, R2, and R3, of rows and with two whole plot treatments, A1 and A2. There are six split plot treatments, T1, T2, T3, T4, T5, and T6 that are in a 6 × 6 Latin square arrangement.

	C1	C2	C3	C4	C5	C6
R1 A1	T1	T2	T3	T4	T5	T6
R1 A2	T2	T3	T4	T5	T6	T1
R2 A1	T3	T4	T5	T6	T1	T2
R2 A2	T4	T5	T6	T1	T2	T3
R3 A1	T5	T6	T1	T2	T3	T4
R3 A2	T6	T1	T2	T3	T4	T5

This design could have appeared in Chapter 3 but it is presented here as an introduction to the more complicated design in Example 5.1.

A key-out of the degrees of freedom for such an arrangement is

Source of variation	Degrees of freedom, example	Degrees of freedom, general
Total	36	k^2
Correction for mean	1	1
Columns	5	$k - 1$
R, pairs of rows	2	$k/a - 1$
A	1	$a - 1$
$A \times R$	2	$(a - 1)(k/a - 1)$
T	5	$k - 1$
$A \times T$	5	$(a - 1)(k - 1)$
Residual = error	15	$(k - 1)(k - 2 - a + 1)$

To form a split block design with split plots as described above, use the columns as well as the rows as whole plots. Such a design is given in the following example.

Example 5.1. To construct a split block experiment design with split plots for the above arrangement, form pairs of columns in the same manner as was done for rows for a second set of whole plot factor B treatments, B1 and B2. The pairs of columns are numbered as C1, C2, and C3. The following is a randomized plan of a split block experiment design with the columns and rows being used as wpeus and with the factor T split plot treatments arranged in a 6 × 6 Latin square design.

	C1		C1		C2		C2		C3		C3	
	B2		B1		B2		B1		B1		B2	
R1 A1	T1	2	T2	5	T4	6	T3	9	T6	8	T5	5
R1 A2	T4	9	T3	7	T6	8	T5	4	T1	3	T2	5
R2 A2	T6	8	T5	5	T1	3	T2	5	T4	9	T3	7
R2 A1	T3	3	T6	4	T5	3	T1	0	T2	1	T4	2
R3 A1	T5	5	T1	5	T2	5	T4	9	T3	7	T6	8
R3 A2	T2	6	T4	8	T3	7	T6	8	T5	6	T1	3

An analysis of variance for the above artificial data set is presented in Table 5.1. By interchanging terms in the model, a Type I analysis often shows the confounding structure better than a Type III analysis does.

To obtain an analysis of the data in Table 5.1, four models are included in the code in Appendix 5.1. These are

$$Y = R\,A\,A * R\,C\,B\,B * C\,T\,A * T,$$
$$Y = R\,A\,A * R\,C\,B\,B * C\,T\,B * T,$$
$$Y = R\,A\,A * R\,C\,B\,B * C\,T\,A * T\,B * T, \text{ and}$$
$$Y = R\,A\,A * R\,C\,B\,B * C\,T\,A * T\,B * T\,A * B * T.$$

Table 5.1. Type I Analysis of Variance.

Source of variation	Degrees of freedom	Sum of squares	Mean square	F-value
Total	36	1298		
Correction for mean	1	1089		
Rows	5	81.67		
R	2	33.50	16.75	1.04
A	1	16.00	16.00	1.00
$A \times R =$ Error A	2	32.17	16.08	
Columns	5	2.67		
C	2	0.50	0.25	1.32
B	1	1.78	1.78	9.37
$B \times C =$ Error B	2	0.39	0.19	
Treatments, T	5	103.33	20.67	29.96
$A \times T$	5	6.52	1.30	1.88
$B \times T$	4	2.05	0.51	0.74
$A \times B \times T$	4	7.96	1.99	2.88
Residual = Error T	7	4.80	0.69	

Type I and Type III sums of squares are equal for the first model demonstrating that the effects in this model are orthogonal. The least square means are estimable. The second model was used to obtain the B*T interaction sum of squares and means. Here again the Type I and Type III sums of squares are the same indicating orthogonality of the effects in the model. For the third model, there are only four degrees of freedom for the B*T interaction instead of five. That is, the A*T and B*T interactions are partially confounded. The least square means are not estimable for this model when using SAS PROC GLM. Additional confounding is introduced in the fourth model as shown in a Type III analysis. Using a series of models such as the above and interchanging the terms in a model is useful to determine which effects are partially or completely confounded with each other. The means and/or solutions for effects may be obtained from the SAS PROC GLM codes given in Appendix 5.1. There are three error terms in the above analysis of variance when the effects of factors A, B, and T are considered to be fixed effects. The A effects are tested with the Error A, the B effects with Error B, and the T and interactions with T effects, with Error C. A fourth error term would be required for testing for the presence of the A × B interaction.

Yates (1933) presents a numerical example for 12 treatments arranged in a 12 × 12 Latin square experiment design. These 12 treatments are formed from combinations of various fertilizer treatments. Two varieties were placed in six pairs of rows in a randomized complete block arrangement. Then, two phosphate treatments were applied to the six pairs of columns to form a randomized complete block arrangement. He discusses various analyses for this split block design with split plots-designed experiments. A design such as this would be useful in investigating a variety of factors, such as density, time of planting, herbicides, fungicides, and so on.

5.3. FACTOR A TREATMENTS ARE THE WHOLE PLOT TREATMENTS AND FACTORS B AND C TREATMENTS ARE IN A SPLIT BLOCK ARRANGEMENT WITHIN EACH WHOLE PLOT

An experiment was laid out with $a = 4$ fertilizer treatments as factor A. A randomized complete block experiment design was used for factor A. Then in each whole plot experimental unit of factor A, $b = 8$ genotypes, factor B, were laid out perpendicular to $c = 4$ herbicide treatments, factor C. There were $r = 5$ complete blocks and hence five randomizations of the four fertilizer treatments, factor A. This results in a three-factor factorial treatment design. There were $ar = 4(5) = 20$ randomizations of the factor B treatments and 20 of the factor C treatments. Note that the speu for factor B is different from the speu for factor C. This design should not be confused with the one in the previous section. A linear model for this experiment design is

$$Y_{ghij} = \mu + \rho_g + \alpha_h + \delta_{gh} + \beta_i + \alpha\beta_{hi} + \pi_{ghi} + \gamma_j + \alpha\gamma_{hj} + \eta_{ghj} + \beta\gamma_{ij} + \alpha\beta\gamma_{hij} + \varepsilon_{ghij}$$

where $g = 1, \ldots, r$, $h = 1, \ldots, a$, $i = 1, \ldots, b$, and $j = 1, \ldots, c$,

μ = general overall mean effect,
ρ_g = gth replicate or block effect identically and independently distributed with mean zero and variance σ_ρ^2,
α_h = effect of hth level of factor A,
δ_{gh} = ghth random error effect identically and independently distributed, IID, with mean zero and variance σ_δ^2,
β_i = effect of ith level of factor B,
$\alpha\beta_{hi}$ = interaction effect hith combination of factors A and B,
π_{ghi} = ghith random error effect IID with mean zero and variance σ_π^2,
γ_j = effect of jth level of factor C,
$\alpha\gamma_{hj}$ = interaction effect of hjth combination of factors A and C,
η_{ghj} = ghjth random error effect IID with zero mean and variance σ_η^2,
$\beta\gamma_{ij}$ = interaction effect of ijth combination of factors B and C,
$\alpha\beta\gamma_{hij}$ = interaction effect of hijth combination of factors A, B, and C, and
ε_{ghij} = ghijth random error effect IID with zero mean and variance σ_ε^2.

The random effects ρ_h, δ_{hi}, π_{ghi}, η_{ghj}, and ε_{ghij} in the above model are assumed to be mutually independent.

A schematic layout for one replicate of this experiment design would be as shown below:

A1				A2				A3				A4			
C1	C2	C3	C4	C1	C2	C3	C4	C1	C2	C3	C4	C1	C2	C3	C4
B1				B1				B1				B1			
B2				B2				B2				B2			
B3				B3				B3				B3			
B4				B4				B4				B4			
B5				B5				B5				B5			
B6				B6				B6				B6			
B7				B7				B7				B7			
B8				B8				B8				B8			

A partitioning of the degrees of freedom for the above in an analysis of variance table for the general case and for $r = 5$, $a = 4$, $b = 8$, and $c = 4$ is presented below:

	Degrees of freedom	
Source of variation	General	Example
Total	$abcr$	$640 = 4(8)(4)(5)$
Correction for mean	1	1
Replicate = R	$r - 1$	4
Factor A, fertilizer, = A	$a - 1$	3
$A \times R$ = error A	$(a-1)(r-1)$	12
Factor B, genotype, = B	$b - 1$	7

(*Continued*)

Source of variation	Degrees of freedom	
	General	Example
$A \times B$	$(a-1)(b-1)$	21
$B \times R$ within A = error B	$a(b-1)(r-1)$	112
Factor C, herbicide, = C	$c-1$	3
$A \times C$	$(a-1)(c-1)$	9
$C \times R$ within A = error C	$a(c-1)(r-1)$	48
$B \times C$	$(b-1)(c-1)$	21
$A \times B \times C$	$(a-1)(b-1)(c-1)$	63
$B \times C \times R$ within A = error ABC	$a(b-1)(c-1)(r-1)$	336

There are four different error terms as there were for the design in the previous section. However, some of the effects have different error terms than for the previous case. Factor B and the $A \times B$ interaction have the same error term for this design. Factor C and the $A \times C$ interaction use another error term, error C, as they have experimental units different from that of factor B. The $B \times C$ and $A \times B \times C$ interactions have the same error term, error ABC. Note that the $B \times C$ interaction is obtained from within each level of factor A and hence will have the same error term as the $A \times B \times C$ interaction.

5.4. FACTORS A AND B IN A STANDARD SPLIT PLOT EXPERIMENT DESIGN AND FACTOR C IN A SPLIT BLOCK ARRANGEMENT OVER BOTH FACTORS A AND B

This type of experiment design has been encountered several times in the course of statistical consulting. The first one of these consisted of $a = 3$ levels of fertilizer for factor A and $b = 7$ genotypes for factor B. Factor C represented three methods of preparing the soil or preconditioning the soil that was done prior to the application of factors A and B. Factor A, the whole plot treatments, and factor B, the split plot treatments, were arranged in a standard split plot experiment design. For $r = 4$ replicates of a randomized complete block design, there were $r = 4$ randomizations of the fertilizer treatments, factor A (F), and $ar = 12$ randomizations of the genotypes, factor B (G). Factor C was in a split block arrangement across the $ab = 21$ *speu*s within each block. There were four randomizations of the three preconditioning treatments, factor C (P). The treatment design structure is a three factor factorial arrangement, denoting the fertilizer treatments as F1, F2, and F3, the genotypes as G1, G2, ..., G7, and the preconditioning soil treatments as P1, P3, and P3. A schematic layout for one replicate of this experiment design is given below:

	F1	F2	F3
P1	G1 G2 G3 G4 G5 G6 G7	G1 G2 G3 G4 G5 G6 G7	G1 G2 G3 G4 G5 G6 G7
P2	G1 G2 G3 G4 G5 G6 G7	G1 G2 G3 G4 G5 G6 G7	G1 G2 G3 G4 G5 G6 G7
P3	G1 G2 G3 G4 G5 G6 G7	G1 G2 G3 G4 G5 G6 G7	G1 G2 G3 G4 G5 G6 G7

A linear model for the above designed experiment is

$$Y_{ghij} = \mu + \rho_g + \alpha_h + \delta_{gh} + \beta_i + \alpha\beta_{hi} + \pi_{ghi} + \gamma_j + \varphi_{gj} + \alpha\gamma_{hj} + \eta_{gij} + \beta\gamma_{ij}$$
$$+ \omega_{gij} + \alpha\beta\gamma_{hij} + \varepsilon_{ghij}$$

where $g = 1, \ldots, r = 4, h = 1, \ldots, a = 3, i = 1, \ldots, b = 7,$ and $j = 1, \ldots, c = 3$,

μ = general overall mean effect,

ρ_g = gth random replicate or block effect identically and independently distributed with mean zero and variance σ_ρ^2,

α_h = effect of hth level of factor A,

δ_{gh} = ghth random error effect identically and independently distributed, IID, with mean zero and variance σ_δ^2,

β_i = effect of ith level of factor B,

$\alpha\beta_{hi}$ = interaction effect hith combination of factors A and B,

π_{ghi} = ghith random error effect IID with mean zero and variance σ_π^2,

γ_j = effect of jth level of factor C,

φ_{gj} = gjth random error effect IID with zero mean and variance σ_φ^2,

$\alpha\gamma_{hj}$ = interaction effect of hjth combination of factors A and C,

η_{ghj} = ghjth random error effect IID with zero mean and variance σ_η^2,

$\beta\gamma_{ij}$ = interaction effect of ijth combination of factors B and C,

ω_{gij} = gijth random error effect IID with zero mean and variance σ_ω^2,

$\alpha\beta\gamma_{hij}$ = interaction effect of hijth combination of factors A, B, and C, and

ε_{ghij} = ghijth random error effect IID with zero mean and variance σ_ε^2.

The random effects ρ_g, δ_{gh}, π_{ghi}, φ_{gj}, η_{ghj}, ω_{gij}, and ε_{ghij} are assumed to be mutually independent.

A partitioning of the degrees of freedom in an analysis of variance is given below:

Source of variation	Degrees of freedom	
	General	Example
Total	abcr	252
Correction for mean	1	1
Replicate = R	$r - 1$	3
Factor $A = A$	$a - 1$	2
$A \times R$ = error A	$(a-1)(r-1)$	6
Factor $B = B$	$b - 1$	6
$A \times B$	$(a-1)(b-1)$	12
$B \times R$ within A = error B	$a(b-1)(r-1)$	54
Factor $C = C$	$c - 1$	2
$C \times R$ = error C	$(c-1)(r-1)$	6
$A \times C$	$(a-1)(c-1)$	4

FACTORS A AND B IN A STANDARD SPLIT PLOT EXPERIMENT DESIGN   129

(*Continued*)

Source of variation	Degrees of freedom	
	General	Example
$A \times C \times R$ = error AC	$(a-1)(c-1)(r-1)$	12
$B \times C$	$(b-1)(c-1)$	12
$B \times C \times R$ = error BC	$(b-1)(c-1)(r-1)$	36
$A \times B \times C$	$(a-1)(b-1)(c-1)$	24
$A \times B \times C \times R$ = error ABC	$(a-1)(b-1)(c-1)(r-1)$	72

There are six error terms in the above analysis of variance table. The design of the experiment is a reasonable one but such a design adds to the complexity of the statistical analysis of the data from the experiment. A design such as the above is useful for obtaining information on more factors and their interactions.

Example 5.2. A numerical example for the above experiment design was constructed using randomly generated numbers. Only two replicates are used rather than four to illustrate the computations and outputs. A SAS PROC GLM code, the data set, and the output are presented in Appendix 5.2. An analysis of variance table and the associated F-tests are given in Table 5.2. From the output, note that Type I and Type III sums of squares are the same, indicating orthogonality of the various effects in the model described above.

From the SAS code given in Appendix 5.2, the six F-statistics given in Table 5.2 were calculated. Least squares means are also obtained for P, F, and G. Means for combinations of P, F, and G may be obtained by inserting the combinations desired in the LSMEANS command.

Table 5.2. Analysis of Variance and Associated F-Statistics for the Data in Appendix 5.2.

Source of variation	Degrees of freedom	Sum of squares	Mean square	F-value
Total	126	4202		
Correction for mean	1	2933.84		
Replicate = R	1	0.79	0.79	
P	2	27.63	13.82	0.69
$P \times R$	2	40.02	20.01	
F	2	31.35	15.67	0.39
$F \times R$	2	80.21	40.10	
$F \times P$	4	46.17	11.54	4.50
$F \times P \times R$	4	10.27	2.57	
G	6	70.38	11.73	1.15
$G \times F$	12	188.76	15.73	1.55
$G \times R$ within F	18	181.33	10.07	
$G \times P$	12	122.48	10.21	1.14
$G \times P \times R$	12	107.65	8.97	
$G \times F \times P$	24	183.38	7.64	1.03
$G \times F \times P \times R$	24	177.73	7.41	

5.5. A COMPLEXLY DESIGNED EXPERIMENT

A researcher designed an experiment using an area consisting of 55 rows with 40 vines per row in a vineyard that was available for experimentation. The following was his description of the experiment. The vineyard was laid out in such a way that there were three rows of vines with their own roots, O, and then two row of vines grafted onto *phyllonera* resistant roots, P. This arrangement for each set of five rows was repeated 11 times, making a systematic arrangement of the two types of roots. Denote the set of five rows of vines, OOOPP, as a column and number the columns from 1 to 11. The first five rows, the middle five rows of vines, and the last five rows (columns 1, 6, and 11) were not used for the factors being investigated. Vines 1, 14, 27, and 40 were not used. The first row of the own rooted vines, O, also was not used. These were considered to be guard rows and vines. A plot consisting of two rows of vines by six vines was the smallest plot size, there being 96 plots in all. Three nitrogen treatments, N1, N2, and N3, were applied to experimental units of size four columns, 2–5 or 7–10, by 12 vines. The three levels of nitrogen were randomly and independently assigned within columns 2–5 and again within columns 7–10. A third factor type of cover had an experimental unit of size two columns by twelve vines. The two types of cover, C1 and C2, are maximum control of weeds and cover crop (minimum competition) and semi-sod cover (maximum competition).The two types of cover were randomly allotted within each nitrogen experimental unit. The fourth factor was pruning severity, P1 and P2. The experimental unit size for the pruning treatments was six vines by two columns. These two treatments were randomly allotted within each type of cover experimental unit. The fifth factor was type of thinning used in the grape clusters, T1 = not thinned and T2 = thinned. The experimental unit for this factor was 12 vines by one column. The two thinning treatments were randomized within the type of cover plots.

A statistical consultant was contacted to determine an appropriate analysis for data from this experiment. A first step for the consultant is to draw a schematic layout for this experiment. Since the layout is somewhat complex, consider only the nitrogen and type of root layout of six experimental units as a first step. Then add the type of cover treatments to obtain the following schematic plan:

	Block 1				Block 2			
	Column				Column			
	2	3	4	5	7	8	9	10
Vine	OOPP	OOPP	OOPP	OOPP	OOPP	OOPP	OOPP	OOPP
2								
3	N1	C1	N1	C2	N1	C1	N1	C2
4								
5								

Row								
6								
7								
8								
9	N1	C1	N1	C2	N1	C1	N1	C2
10								
11								
12								
13								
15								
16								
17	N2	C1	N2	C2	N2	C1	N2	C2
18								
19								
20								
21								
22								
23	N2	C1	N2	C2	N2	C1	N2	C2
24								
25								
26								
28								
29								
30	N3	C1	N3	C2	N3	C1	N3	C2
31								
32								
33								
34								
35								
36	N3	C1	N3	C2	N3	C1	N3	C2
37								
38								
39								

Without delving further into the design structure for this experiment, what is the statistical analysis so far? Note that nitrogen and root type are in a split block arrangement and that cover type is in a split plot arrangement with nitrogen as the whole plot. The experiment design is similar to the one in the previous section. The response variable was yield of grapes. Considering only the two types of roots and using all eleven columns, the following partitioning of degrees of freedom is obtained for an analysis of variance table:

Source of variation	Degrees of freedom
Total	22
Correction for the mean	1
Columns	10
Root type, O vs. P	1
Root type × column	10

Since the root type treatments were not randomized, no error term is available. However, we know that the systematic arrangement tends to bias the Root type × column error mean square upward, and thus an F-test may still be made keeping in mind that the Type I error is biased upward.

For the schematic arrangement given above, a partitioning of the degrees of freedom in an analysis of variance is:

Source of variation	Degrees of freedom	
Total	$24 = 2(3)(2)(2)$	
Correction for the mean	1	
Block (columns 2–5 vs. 7–10)	1	
Root (O vs. P)	1	
Root × block	1	error mean square for root
Nitrogen	2	
Nitrogen × block	2	error mean square for nitrogen
Root × nitrogen	2	
Root × nitrogen × block	2	error mean square for interaction
Cover (C1 vs. C2)	1	
Cover × nitrogen	2	
Cover × block within nitrogen	3	error for cover and cover × nitrogen
Cover × root	1	
Cover × root × block	1	error for cover × root
Cover × nitrogen × root	2	
Cover × nitrogen × root × block	2	error for cover × nitrogen × root

This analysis of variance is similar to the one in the previous section. The number of degrees of freedom associated with the various error terms is only 1 or 2.

A COMPLEXLY DESIGNED EXPERIMENT   133

The analyst needs to consider pooling some of the error mean squares and perhaps even the root × column mean square from the 11 columns. One of the pooling procedures described by Bozivich et al. (1956) may be of use here. Another procedure would be to partition the nitrogen sum of squares into single degree of freedom contrasts and likewise for all interactions with nitrogen. A half normal probability plot of the 23 (omitting the one degree of freedom for the correction for the mean) single degree of freedom contrasts could then be made and a search for outlying observations made. An error mean square from the remaining observations could then be obtained. When there is a paucity of degrees of freedom for an error term, the analyst must resort to other means, perhaps even obtaining an estimate of the error mean square from other experiments or analyses.

A schematic layout for the pruning and thinning treatments for columns 2 to 5 and vines 2 to 13 is presented below:

	Column			
	2	3	4	5
Vine	O O P P	O O P P	O O P P	O O P P
2	N1 C1	N1 C1	N1 C2	N1 C2
3				
4	P1 T1	P2 T1	P1 T1	P2 T1
5				
6				
7				
8	N1 C1	N1 C1	N1 C2	N1 C2
9				
10	P1 T2	P2 T2	P1 T2	P2 T2
11				
12				
13				

The pruning treatments, P1 and P2, and the thinning treatments, T1 and T2, are in a split block arrangement with each other and both are in a split plot arrangement to cover type. There are 12 randomizations for thinning treatments. Also, there are 12 randomizations for pruning treatments. The experimental units for these two types of treatments are different and hence they have different error terms. The error term for their interaction and the terms involving this interaction are also different owing to the split block arrangement of the thinning and the pruning treatments. Thus, the partitioning of degrees of freedom in an analysis of variance table is as follows:

Source of variation	Degrees of freedom	
Total	96	
Correction for the mean	1	
Block (columns 2–5 vs. 7–10)	1	
Root (O vs. P)	1	
Root × block	1	error for root
Nitrogen	2	
Nitrogen × block	2	error for nitrogen
Nitrogen × root	2	
Nitrogen × root × block	2	error for nitrogen × root
Cover (C1 vs. C2)	1	
Cover × nitrogen	2	
Cover × block within nitrogen	2	error for cover and cover × nitrogen
Cover × root	1	
Cover × root × block	1	error for cover × root
Cover × root × nitrogen	2	
Cover × root × nitrogen × block	2	error for cover × root × nitrogen
Pruning (P1 vs. P2)	1	
Pruning × cover	1	
Pruning × nitrogen	2	
Pruning × cover × nitrogen	2	
Pruning × root	1	
Pruning × root × cover	1	
Pruning × root × nitrogen	2	
Pruning × root × cover × nitrogen	2	
Pruning × block within cover, root, and nitrogen	12	error for pruning and interactions with pruning
Thinning (T1 vs. T2)	1	
Thinning × root	1	
Thinning × nitrogen	2	
Thinning × cover	1	
Thinning × nitrogen × root	2	
Thinning × cover × root	1	
Thinning × nitrogen × cover	2	
Thinning × nitrogen × cover × root	2	
Thinning × block within root, nitrogen, and cover	12	error for thinning and interactions with thinning
Thinning × pruning	1	
Thinning × pruning × root	1	
Thinning × pruning × nitrogen	2	
Thinning × pruning × cover	1	
Thinning × pruning × root × cover	1	
Thinning × pruning × root × nitrogen	2	
Thinning × pruning × nitrogen × cover	2	
Thinning × pruning × root × nitrogen × cover	2	
Thinning × pruning × block within root, nitrogen, and cover	12	error for thinning × pruning and other interactions with this interaction

SOME RULES TO FOLLOW FOR FINDING AN ANALYSIS   135

There are nine different error terms in the above analysis of variance. Laying out experiments the way its been done in this chapter makes perfectly good sense to an experimenter and uses the experimental material available for an experiment, even if the analysis presents somewhat of a nightmare for the analyst.

5.6. SOME RULES TO FOLLOW FOR FINDING AN ANALYSIS FOR COMPLEXLY DESIGNED EXPERIMENTS

The following set of rules and algorithms were set forth by Federer (1975, 1977) as aids for the statistical analyst facing the task of determining an appropriate statistical analysis for complexly designed experiments.

Rule I: Make no assumptions about the form of the statistical design; always determine the exact experimental procedure, not the stated one. (The experimenter stated that the design in the previous section was a five factor factorial treatment design laid out as a randomized complete block experiment design.)

Rule II: Determine the experimental unit for levels of each category (factor, block, etc.), then determine any common experimental units for combinations for all possible pairs of categories, then for all possible triplets, etc.

Rule III: Count the number of randomizations for each category (factor, block, etc.) in the experiment, then count number of randomizations for combinations of all possible pairs of categories, then for all possible triplets of categories, etc.

Rule IV: Determine which category levels are nested within another category level and which are crossed.

Rule V: Ignore complexity of design in first keying out of degrees of freedom; relate key-out to nearest known design.

Algorithm I: Keying out degrees of freedom in the ANOVA.

1. At every step, perform the simplest key-out of degrees of freedom that is possible.
2. First determine total degrees of freedom and partition into one for the correction term and the rest for the remainder of the sum of squares corrected for the mean.
3. Key-out degrees of freedom for the category or categories offering the least difficulty.
4. Key-out degrees of freedom for all possible pairs of categories, then all possible triplets, and so on, excluding any pairs, triplets, and so on, not needed.

5. Isolate all degrees of freedom in the ANOVA for which the partitioning is not understood.
6. Defer the partitioning of degrees of freedom that are not completely understood.
7. Approach the partitioning in step 6 from different directions in order to reduce steps 5 and 6 to the null set. Note that further partitioning may be impossible until more information is available.

Rule VI: No computations of sums of squares should be performed until the correctness of the degrees of freedom key-out in the ANOVA has been ascertained and the appropriate error variances have been designated.

Rule VII: With almost probability one, experiments and surveys involving humans and animals will have effects completely or partially confounded and one will need to follow rules I through V in order to ascertain this.

Rule VIII: Be prepared to spend a considerable amount of time and effort in unraveling the confounding schemes in any human or animal experiment as planned by the researcher (and perhaps even by a statistician).

Algorithm II. Computing sums of squares in the ANOVA.

1. At every step compute the simplest ANOVA sums of squares, that is, sums of squares assuming nesting even though there was no nesting.
2. Compute sums of squares for degrees of freedom key-outs in steps 2, 3, and 4 of algorithm I. For many investigations this may be a desk calculator job.
3. For partially confounded effects, it may be necessary to solve a set of normal equations prior to computing the sums of squares if needed software is unavailable.
4. If steps 5 and 6 of algorithm I have not been reduced to the null set, nothing should be done about a further partitioning of the sums of squares.

Algorithm III: Determining error variances for F-tests.

1. Factors with the same type of experimental units *may* have the same error variances.
2. Factors with different experimental units almost always have different error variances.
3. In order to check the validity of the error variance, determine the appropriate error variance assuming other effects are absent from the experiment for single factors, then for pairs of factors, etc.
4. Check to determine if partially confounded effects can be estimated from two sources and with two different error variances.
5. Check your decisions with known situations if possible.

SOME RULES TO FOLLOW FOR FINDING AN ANALYSIS

Some rules for partitioning out the degrees of freedom in an analysis of variance for complexly designed experiments have been presented by Federer (1977). These are given below.

Rule 1: It is essential to determine precisely the experimental unit for each treatment (factor) and to define the sampling structure of the experimental unit for each factor separately as well as jointly.

One way of implementing this rule is by drawing a schematic sketch of the experiment as laid out and by determining if anything occurred during the course of the experiment that is part of the design of the experiment and the response model. Just like the example in the previous section, it may be necessary to draw more than one sketch of the experimental layout. For spatially or temporally designed experiments, this usually presents little difficulty. It is much more difficult to do this with experiments on animals, humans, and other situations such as may be encountered in ecological experiments.

Rule 2: It is essential to determine precisely what the sampling and experimental procedure actually is rather than what it is purported to be.

During the course of statistical consulting, the consultee has often stated that they used a randomized complete block design when in fact the design was a split split plot, a split block, or other design. Also, many consultees are unclear as to the meaning of replication and often confuse observational and sampling units with experimental units. The actual layout and conduct of the experiment must be known before one can determine the type of design used. This is why the experimental unit sizes were obtained from the experimenter for the example in the previous section.

Rule 3: It is essential to know the number of randomizations used for each factor singly, for pairs of factors, for triplets of factors, and so forth. If the number of randomizations differs, one can expect different sizes of experimental units and perhaps some confounding of factor effects. The number of randomizations used for each factor needs to be known. If the number of randomizations differs, the size of the experimental units will differ and different error mean squares will be required for testing effects.

Rule 4: It is essential to determine if levels of factors are crossed or nested within levels of other factors.

Rule 5: An error variance for a treatment contrast of levels of a factor is determined from the variance among experimental units treated alike (i.e., same treatment) where the experimental units are those of the factor involved. Note that there may be several error variances in an experiment.

Rule 6: It is essential to have a complete and meaningful key-out of the degrees of freedom in an analysis of variance table prior to any statistical calculations.

Rule 7: It is desirable to recheck all assumptions involved in the statistical analysis such as independence of observations, additivity of effects, equality of components of a mean square, and normality of residuals, to ascertain that they are satisfactory for the material used in the experiment under consideration.

Rule 8: After computing means and effects, it is highly desirable to study residuals prior to considering the sums of squares and mean squares.

Rule 9: It is very useful to prepare graphical displays of the data, the means, and the residuals.

Rule 10: When fully satisfied with the preceding nine considerations, compute the sums of squares and the mean squares and complete the statistical calculations including tests of significance.

5.7. COMMENTS

As may be observed from the above, experiment designs can become rather complex and this may call for a complex statistical analysis of the data from such an experiment. If the analyst follows the rules and suggestions outlined in the previous section, it should be possible to obtain an appropriate statistical analysis for the data from an experiment. The following chapter presents an even more complex situation than the one above. Such experiments tax the creativity and ingenuity of the analyst.

Most books and courses on statistical methods for the analysis of data from experiments never get beyond one error term per experiment. The above examples demonstrate the need for more than one error term per experiment. When presented with the above analyses, the experimenters were surprised if not shocked at the nature of the analysis for their experiments. For the reader, the above examples came from real experiments that add reality to the need for discussing such situations.

An analyst may consider using SAS PROC MIXED to take the random effects into account. If the effects in the model are orthogonal, nothing is to be gained using this code as the means will be identical. Owing to the nature and subtleties of using mixed effects analyses, it may be wise to forgo using this procedure for data from experiments designed in the fashion of this chapter. The analyst may also consider using a multiple comparisons procedure for comparing pairs of means. For such experiment designs as discussed in this chapter and in Chapters 3 and 4, there will be numerous standard errors of a difference between pairs of means. This means there will be numerous values of the lsd, the hsd, and others. If the analyst wishes to compare a pair of means using a particular multiple comparisons procedure, it will be necessary to first find the appropriate standard error of the difference of the pair of means involved and then compute the appropriate value of the lsd, hsd, and so forth.

5.8. PROBLEMS

Problem 5.1. Use the command "/solution" in the model statement for the models with two and three factor interactions in Example 5.1. Determine which means and differences of effects are estimable. (Note that SAS subtracts the highest numbered effect from all effects of a variable to make the highest numbered effect equal to zero. The standard errors given are standard errors of a difference between two effects, the effect and the highest numbered effect.)

Problem 5.2. For the data given by Yates (1933, page 22) for the split block experiment design with split plot treatments in a 12 × 12 Latin square arrangement, obtain the analyses for the four models given by the code in Appendix 5.1. Are all means and differences of effects estimable for each of the models?

Problem 5.3. What type of confounding results are appropriate for obtaining the results in Table 5.1 when the A × B interaction term is added to the model?

Problem 5.4. Write an appropriate SAS code for obtaining the A*B interaction and the associated F-test for the data presented in Example 5.1.

5.9. REFERENCES

Bozivich, H., T. A. Bancroft, and H. O. Hartley (1956). Power of analysis of variance test procedures for certain incompletely specified models. Annals of Mathematical Statistics 27:1017–1043.

Federer, W. T. (1975). The misunderstood split plot. In *Applied Statistics* (Editor: R. P. Gupta), North Holland Publishing Company, Amsterdam, pp. 9–38.

Federer, W. T. (1977). Sampling, blocking, and model considerations for split plot and split block designs. Biometrical Journal 19(3):181–200.

Federer, W. T. (1984). Principles of statistical design with special reference to experiment and treatment design. In *Statistics: An Appraisal* (Editors: H. A. and H. T. David), The Iowa State University Press, Ames, Iowa, pp. 77–104.

Federer, W. T. (1991). *Statistics and Society: Data Collection and Interpretation*, Second Edition. Marcel Dekker, Inc., New York, page 174, pp. 1–578.

Yates, F. (1933). The principles of orthogonality and confounding in replicated experiments. The Journal of Agricultural Science XXIII:108–145.

APPENDIX 5.1. EXAMPLE 5.1 CODE

An abbreviated data set, the code, and output from running the code for Example 5.1 are presented below. The complete data set is given on the book's FTP site.

```
data sbsp;
input R A C B T Y;
```

```
/*R is row pair, A if factor A, C is column pair, B is factor B, T is
factor T as described in Example 5.1, Y is the response*/
datalines;
1  1  1  2  1  2
1  1  1  1  2  5
1  1  2  2  4  6
...
3  2  3  1  5  6
3  2  3  2  1  3
;
Proc glm data = sbsp;
class R A C B T;
model Y = R A R*A C B C*B T B*T;
lsmeans A B T B*T;
run;
Proc glm data = sbsp;
class R A C B T;
model Y = R A A*R C B C*B T A*T;
lsmeans A B T A*T;
run;
Proc glm data = sbsp;
class R A C B T;
model Y = R A A*R C B B*C T A*T B*T;
run;
Proc glm data = sbsp;
class R A C B T;
model Y = R A A*R C B C*B T A*T B*T A*B*T;
run;
```

An abbreviated form of the output for the above code and data set is given below:

The GLM Procedure

Dependent Variable: Y

Source	DF	Sum of Squares	Mean Square	F Value	Pr > F
Model	20	193.5833333	9.6791667	9.42	<.0001
Error	15	15.4166667	1.0277778		
Corrected Total	35	209.0000000			

R-Square	Coeff Var	Root MSE	Y Mean
0.926236	18.43261	1.013794	5.500000

Source	DF	Type I SS	Mean Square	F Value	Pr > F
R	2	33.5000000	16.7500000	16.30	0.0002
A	1	16.0000000	16.0000000	15.57	0.0013
R*A	2	32.1666667	16.0833333	15.65	0.0002

APPENDIX

C	2	0.5000000	0.2500000	0.24	0.7871
B	1	1.7777778	1.7777778	1.73	0.2082
C*B	2	0.3888889	0.1944444	0.19	0.8296
T	5	103.3333333	20.6666667	20.11	<.0001
B*T	5	5.9166667	1.1833333	1.15	0.3770

Source	DF	Type III SS	Mean Square	F Value	Pr > F
R	2	22.4508547	11.2254274	10.92	0.0012
A	1	15.0012626	15.0012626	14.60	0.0017
R*A	2	27.1638138	13.5819069	13.21	0.0005
C	2	0.5000000	0.2500000	0.24	0.7871
B	1	1.7777778	1.7777778	1.73	0.2082
C*B	2	0.3888889	0.1944444	0.19	0.8296
T	5	103.3333333	20.6666667	20.11	<.0001
B*T	5	5.9166667	1.1833333	1.15	0.3770

Least Squares Means

A	Y LSMEAN
1	4.74305556
2	6.25694444

B	Y LSMEAN
1	5.72222222
2	5.27777778

T	Y LSMEAN
1	2.66666667
2	4.50000000
3	6.66666667
4	7.16666667
5	4.66666667
6	7.33333333

B	T	Y LSMEAN
1	1	3.43055556
1	2	4.32638889
1	3	7.18055556
1	4	7.86805556
1	5	4.24305556
1	6	7.28472222
2	1	1.90277778
2	2	4.67361111
2	3	6.15277778
2	4	6.46527778
2	5	5.09027778
2	6	7.38194444

The GLM Procedure

Dependent Variable: Y

Source	DF	Sum of Squares	Mean Square	F Value	Pr > F
Model	20	194.1875000	9.7093750	9.83	<.0001
Error	15	14.8125000	0.9875000		
Corrected Total	35	209.0000000			

R-Square	Coeff Var	Root MSE	Y Mean
0.929127	18.06782	0.993730	5.500000

Source	DF	Type I SS	Mean Square	F Value	Pr > F
R	2	33.5000000	16.7500000	16.96	0.0001
A	1	16.0000000	16.0000000	16.20	0.0011
R*A	2	32.1666667	16.0833333	16.29	0.0002
C	2	0.5000000	0.2500000	0.25	0.7796
B	1	1.7777778	1.7777778	1.80	0.1996
C*B	2	0.3888889	0.1944444	0.20	0.8234
T	5	103.3333333	20.6666667	20.93	<.0001
A*T	5	6.5208333	1.3041667	1.32	0.3079

Type III sums of squares are the same as for Type I.

Least Squares Means

A	Y LSMEAN
1	4.83333333
2	6.16666667

B	Y LSMEAN
1	5.76041667
2	5.23958333

T	Y LSMEAN
1	2.66666667
2	4.50000000
3	6.66666667
4	7.16666667
5	4.66666667
6	7.33333333

A	T	Y LSMEAN
1	1	2.25000000
1	2	3.55208333
1	3	6.25000000
1	4	5.65625000

APPENDIX

```
            1    5    4.59375000
            1    6    6.69791667
            2    1    3.08333333
            2    2    5.44791667
            2    3    7.08333333
            2    4    8.67708333
            2    5    4.73958333
            2    6    7.96875000
```

The GLM Procedure

Dependent Variable: Y

Source	DF	Sum of Squares	Mean Square	F Value	Pr > F
Model	24	196.2378788	8.1765783	7.05	0.0009
Error	11	12.7621212	1.1601928		
Corrected Total	35	209.0000000			

R-Square	Coeff Var	Root MSE	Y Mean
0.938937	19.58405	1.077122	5.500000

Source	DF	Type I SS	Mean Square	F Value	Pr > F
R	2	33.5000000	16.7500000	14.44	0.0008
A	1	16.0000000	16.0000000	13.79	0.0034
R*A	2	32.1666667	16.0833333	13.86	0.0010
C	2	0.5000000	0.2500000	0.22	0.8095
B	1	1.7777778	1.7777778	1.53	0.2415
C*B	2	0.3888889	0.1944444	0.17	0.8478
T	5	103.3333333	20.6666667	17.81	<.0001
A*T	5	6.5208333	1.3041667	1.12	0.4027
B*T	4	2.0503788	0.5125947	0.44	0.7762

Source	DF	Type III SS	Mean Square	F Value	Pr > F
R	2	22.1855291	11.0927646	9.56	0.0039
A	1	15.1853369	15.1853369	13.09	0.0040
R*A	2	27.4263657	13.7131829	11.82	0.0018
C	2	1.0097187	0.5048594	0.44	0.6578
B	1	1.7922201	1.7922201	1.54	0.2398
C*B	2	0.0848694	0.0424347	0.04	0.9642
T	5	103.3333333	20.6666667	17.81	<.0001
A*T	4	2.6545455	0.6636364	0.57	0.6887
B*T	4	2.0503788	0.5125947	0.44	0.7762

The GLM Procedure

Dependent Variable: Y

Source	DF	Sum of Squares	Mean Square	F Value	Pr > F
Model	28	204.2000000	7.2928571	10.64	0.0017

Error	7	4.8000000	0.6857143
Corrected Total	35	209.0000000	

R-Square	Coeff Var	Root MSE	Y Mean
0.977033	15.05598	0.828079	5.500000

Source	DF	Type I SS	Mean Square	F Value	Pr > F
R	2	33.5000000	16.7500000	24.43	0.0007
A	1	16.0000000	16.0000000	23.33	0.0019
R*A	2	32.1666667	16.0833333	23.45	0.0008
C	2	0.5000000	0.2500000	0.36	0.7069
B	1	1.7777778	1.7777778	2.59	0.1514
C*B	2	0.3888889	0.1944444	0.28	0.7613
T	5	103.3333333	20.6666667	30.14	0.0001
A*T	5	6.5208333	1.3041667	1.90	0.2123
B*T	4	2.0503788	0.5125947	0.75	0.5896
A*B*T	4	7.9621212	1.9905303	2.90	0.1039

Source	DF	Type III SS	Mean Square	F Value	Pr > F
R	2	28.1123834	14.0561917	20.50	0.0012
A	1	14.6550611	14.6550611	21.37	0.0024
R*A	1	2.0173729	2.0173729	2.94	0.1300
C	2	0.4711620	0.2355810	0.34	0.7206
B	1	1.7693846	1.7693846	2.58	0.1522
C*B	1	0.6444915	0.6444915	0.94	0.3646
T	5	103.8149128	20.7629826	30.28	0.0001
A*T	4	2.9513491	0.7378373	1.08	0.4359
B*T	4	3.5529621	0.8882405	1.30	0.3580
A*B*T	4	7.9621212	1.9905303	2.90	0.1039

APPENDIX 5.2. EXAMPLE 5.2 DATA SET, CODE, AND OUTPUT

The complete data set is given on the book's FTP site.

```
data example;
input R P F G Y;
/*R is replicate, P is factor A, F is factor B, G is genotype, Y is the
response*/
datalines;
1 1 1 1 5
1 1 1 2 9
1 1 1 3 3
1 1 1 4 9
1 1 1 5 1
1 1 1 6 5
1 1 1 7 8
...
2 3 3 5 4
```

APPENDIX

```
2 3 3 6 3
2 3 3 7 9
;
proc glm data = example;
class R P F G;
model Y = R P R*P F F*R P*F P*F*R G F*G G*R(F) G*P G*P*R G*F*P G*F*P*R;
test H = P E = R*P; test H = F E = F*R;
test H = P*F E = P*F*R; test H = G E = G*R(F);
test H = F*G E = G*R(F); test H = G*P E = G*P*R;
test H = G*F*P E = G*F*P*R;
lsmeans P F G;
RUN;
```

The abbreviated output for the above code is presented below:

Dependent Variable: Y

Source	DF	Sum of Squares	Mean Square	F Value	Pr > F
Model	125	1268.158730	10.145270	.	.
Error	0	0.000000	.		
Corrected Total	125	1268.158730			

R-Square	Coeff Var	Root MSE	Y Mean
1.000000	.	.	4.825397

Source	DF	Type I SS	Mean Square	F Value	Pr > F
R	1	0.7936508	0.7936508	.	.
P	2	27.6349206	13.8174603	.	.
R*P	2	40.0158730	20.0079365	.	.
F	2	31.3492063	15.6746032	.	.
R*F	2	80.2063492	40.1031746	.	.
P*F	4	46.1746032	11.5436508	.	.
R*P*F	4	10.2698413	2.5674603	.	.
G	6	70.3809524	11.7301587	.	.
F*G	12	188.7619048	15.7301587	.	.
R*G(F)	18	181.3333333	10.0740741	.	.
P*G	12	122.4761905	10.2063492	.	.
R*P*G	12	107.6507937	8.9708995	.	.
P*F*G	24	183.3809524	7.6408730	.	.
R*P*F*G	24	177.7301587	7.4054233	.	.

Dependent Variable: Y

Tests of Hypotheses Using the Type III MS for R*P as an Error Term

Source	DF	Type III SS	Mean Square	F Value	Pr > F
P	2	27.63492063	13.81746032	0.69	0.5915

Tests of Hypotheses Using the Type III MS for R*F as an Error Term

Source	DF	Type III SS	Mean Square	F Value	Pr > F
F	2	31.34920635	15.67460317	0.39	0.7190

Tests of Hypotheses Using the Type III MS for R*P*F as an Error Term

Source	DF	Type III SS	Mean Square	F Value	Pr > F
P*F	4	46.17460317	11.54365079	4.50	0.0873

Tests of Hypotheses Using the Type III MS for R*G(F) as an Error Term

Source	DF	Type III SS	Mean Square	F Value	Pr > F
G	6	70.3809524	11.7301587	1.15	0.3912
F*G	12	188.7619048	15.7301587	1.55	0.2309

Tests of Hypotheses Using the Type III MS for R*P*G as an Error Term

Source	DF	Type III SS	Mean Square	F Value	Pr > F
P*G	12	122.4761905	10.2063492	1.14	0.4134

Tests of Hypotheses Using the Type III MS for R*P*F*G as an Error Term

Source	DF	Type III SS	Mean Square	F Value	Pr > F
P*F*G	24	183.3809524	7.6408730	1.03	0.4698

Least Squares Means

P	Y LSMEAN
1	4.16666667
2	5.21428571
3	5.09523810

F	Y LSMEAN
1	4.30952381
2	4.66666667
3	5.50000000

G	Y LSMEAN
1	4.77777778
2	5.94444444
3	3.94444444
4	5.11111111
5	3.66666667
6	4.83333333
7	5.50000000

CHAPTER 6

World Records for the Largest Analysis of Variance Table (259 Lines) and for the Most Error Terms (62) in One Analysis of Variance

6.1. INTRODUCTION

The experiment described in this section produced data that resulted in the largest analysis of variance table ever encountered in practice or seen in published literature. Hence, it is claimed that this is a world record for the largest analysis of variance table. There are 259 lines in the analysis of variance table developed for data from the experiment. This was been submitted to Guinness Worlds Records for a world record. Accompanying this table are 62 error terms for the various effects in the experiment. This is claimed to be a record for the number of error terms in one analysis of variance table. This claim has also been submitted to Guinness Worlds Records as a world record. The records were not granted as Guinness did not have a category for items of this nature as they are not general enough and too specific for their more general records.

In Chapters 3, 4, and 5, variations of split plot and split block experiment designs have been discussed. The desires and goals of the experimenter often dictate the experiment design used for each experiment. Some designs may need to be unorthodox in order to attain the goals. In this chapter, a very complexly designed experiment was conducted in the early 1950s in Hawaii. Several departments of the Pineapple Research Institute were involved. Owing to the size and complexity of the experiment, it was dubbed their "Manhattan Project." Federer and Farden (1955)

Variations on Split Plot and Split Block Experiment Designs, by Walter T. Federer and Freedom King
Copyright © 2007 John Wiley & Sons, Inc.

presented one analysis for the data from the experiment. The analysis presented herein differs in some respects from theirs, in that several approaches are possible. To obtain an analysis of data from such an experiment, the procedures described in Chapter 5 have been used.

The design and the field layout of the experiment are described in Section 6.2. In Section 6.3, some preliminary analyses are performed to obtain a starting point for obtaining an analysis of the data from this experiment. Owing to the complexity of the design of the experiment, several approaches are used as guides to the final analysis as given in Section 6.4.

6.2. DESCRIPTION OF THE EXPERIMENT

The continuous production of a single crop on a piece of land for a long period of time might be expected to produce changes in the soil peculiar to the crop grown. Two adjacent pieces of land, one of which was farmed for many years to pineapple and the other to sugar cane, might be appreciably different in soil properties and hence require different management practices for pineapple production. Changes in soil pH and pest levels might be expected. Two adjacent pieces of land with these conditions became available for use by the Pineapple Research Institute (PRI). It was expected that the experiment would continue for several cycles of pineapple production.

The objectives of the experiment were to study long-term effects on soil properties, on pineapple growth, and fruit production, resulting from previous cropping to sugar cane as compared to pineapple. It was designed to study relative responses to fumigation and fertilizer variables as they occur at present and after adjustment for pH, and to investigate the best agronomic practices for pineapple production on lands previously cropped to sugar cane.

Preexperiment fumigation and paper laying were started on 10/24/52 and completed on 11/11/52 (Federer and Farden, 1955). Planting of pineapple slips started in the field on 11/1/52 and was completed on 11/11/52. The plant slips were planted fourteen inches apart. The plots were each one hundred feet long and three field blocks wide. A description of the treatments used follows. As shown in Figure 6.1 in Appendix 6.1, the two pieces of land or types of soil were initially subdivided into eight plots on old sugar cane land and eight plots on old pineapple land. The symbols C, A, P, and L are defined as:

C – old sugar cane land, pH unadjusted

A – old sugar cane land, pH adjusted, sulfur added so that the pH is approximately the same as in P,

P – old pineapple land, pH unadjusted

L – old pineapple land, pH adjusted, and limed so that the pH is approximately the same as in C.

DESCRIPTION OF THE EXPERIMENT   **149**

The preplanting fumigation treatments (See Figure 6.2 for layout) were as follows:

   I – 800 pounds of DD
   II – 400 pounds of DD
   III – no preplanting fumigant.

The postplanting fumigation treatments (See Figure 6.3 for layout) were as follows:

   Y – 40 gallons of 10% EDB
   Z – no postplanting fumigant.

The postplanting fertilizer treatments (See Figure 6.4 for layout) were as follows:

   S – Ammonium sulfate
   U – Urea.

The preplanting fertilizer treatments (See Figure 6.5 for layout) were as follows:

   1. Major nutrients (N, P, K, Mg)
   2. Major nutrients plus minor nutrients (Cu, B, Mo)
   3. Micro nutrients (Cu, B, Mo) alone
   4. No plant fertilizer.

N was applied at the rate of 100 pounds per acre as ammonium sulfate, P at the rate of 200 pounds per acre as super phosphate, K at 100 pounds per acre as potassium sulfate, Mg at 1/20th of the available calcium as sulfate, Cu at 15 pounds per acre as $CuSO_4$, B at 20 pounds per acre as borax, and Mo at one pound per acre as molybdic acid. Treatments 1 and 4 were named as major element treatments, and treatments 2 and 3 were called minor element treatments.

The field design of the experiment is given in Table 6.1 and in Figures 6.1–6.6. The last figure combined with Figure 6.1 gives the layout of the experiment. Considering old sugar cane land and old pineapple land as whole plots, C and A are split plot treatments to the former and P and L are split plot treatments to old pineapple land. Considering the treatment variables applied to rows, preplanting fumigation is a whole plot treatment, postplanting fumigation is a split plot treatment, and postplanting fertilizer treatments is a split split plot with rows as the blocks. Preplanting fertilizer treatments 1 and 4 (major elements) and 2 and 3 (minor elements) are in a split plot arrangement to soils but a split block arrangement with respect to C and A and also P and L, and 1 and 4 are split split plots as are 2 and 3 to soils but split block to C and A and also P and L. The latter design in rows is in a split block arrangement to the experiment design in columns. The rules for unraveling the degrees of freedom in an analysis

Table 6.1. Field Plan for Sugar Cane Land, C and A, and Pineapple Land, P and L Experiment.

Replicate 1 or row 1 (last number in a cell)

Old cane land   Old pineapple land

CI	CI	CI	CI	AI	AI	AI	AI	LI	LI	LI	LI	PI	PI	PI	PI
Y1	Y4	Y2	Y3	Y2	Y3	Y1	Y4	Y4	Y1	Y3	Y2	Y3	Y2	Y1	Y4
CI	CI	CI	CI	AI	AI	AI	AI	LI	LI	LI	LI	PI	PI	PI	PI
Y1	Y4	Y2	Y3	Y2	Y3	Y1	Y4	Y4	Y1	Y3	Y2	Y3	Y2	Y1	Y4
CI	CI	CI	CI	AI	AI	AI	AI	LI	LI	LI	LI	PI	PI	PI	PI
Z1	Z4	Z2	Z3	Z2	Z3	Z1	Z4	Z4	Z1	Z3	Z2	Z3	Z2	Z1	Z4
CI	CI	CI	CI	AI	AI	AI	AI	LI	LI	LI	LI	PI	PI	PI	PI
Z1	Z4	Z2	Z3	Z2	Z3	Z1	Z4	Z4	Z1	Z3	Z2	Z3	Z2	Z1	Z4
CII	CII	CII	CII	AII	AII	AII	AII	LII	LII	LII	LII	PII	PII	PII	PII
Z4	Z1	Z3	Z2	Z2	Z3	Z1	Z4	Z4	Z1	Z2	Z3	Z3	Z2	Z4	Z1
CII	CII	CII	CII	AII	AII	AII	AII	LII	LII	LII	LII	PII	PII	PII	PII
Z4	Z1	Z3	Z2	Z2	Z3	Z1	Z4	Z4	Z1	Z2	Z3	Z3	Z2	Z4	Z1
CII	CII	CII	CII	AII	AII	AII	AII	LII	LII	LII	LII	PII	PII	PII	PII
Y4	Y1	Y3	Y2	Y2	Y3	Y1	Y4	Y4	Y1	Y2	Y3	Y3	Y2	Y4	Y1
CII	CII	CII	CII	AII	AII	AII	AII	LII	LII	LII	LII	PII	PII	PII	PII
Y4	Y1	Y3	Y2	Y2	Y3	Y1	Y4	Y4	Y1	Y2	Y3	Y3	Y2	Y4	Y1
CIII	CIII	CIII	CIII	AIII	AIII	AIII	AIII	LIII	LIII	LIII	LIII	PIII	PIII	PIII	PIII
Y4	Y1	Y2	Y3	Y3	Y2	Y1	Y4	Y1	Y4	Y3	Y2	Y3	Y2	Y4	Y1
U1	U1	U1	U1	U1	U1	U1	U1	U1	U1	U1	U1	U1	U1	U1	U1
CIII	CIII	CIII	CIII	AIII	AIII	AIII	AIII	LIII	LIII	LIII	LIII	PIII	PIII	PIII	PIII
Y4	Y1	Y2	Y3	Y3	Y2	Y1	Y4	Y1	Y4	Y3	Y2	Y3	Y2	Y4	Y1
CIII	CIII	CIII	CIII	AIII	AIII	AIII	AIII	LIII	LIII	LIII	LIII	PIII	PIII	PIII	PIII
Z4	Z1	Z2	Z3	Z3	Z2	Z1	Z4	Z1	Z4	Z3	Z2	Z3	Z2	Z4	Z1
CIII	CIII	CIII	CIII	AIII	AIII	AIII	AIII	LIII	LIII	LIII	LIII	PIII	PIII	PIII	PIII
Z4	Z1	Z2	Z3	Z3	Z2	Z1	Z4	Z1	Z4	Z3	Z2	Z3	Z2	Z4	Z1

DESCRIPTION OF THE EXPERIMENT 151

Table 6.1. (*Continued*)

Replicate 2 or row 2

Old cane land   Old pineapple land

AII	AII	AII	AII	CII	CII	CII	CII	PII	PII	PII	PII	LII	LII	LII	LII
Y1	Y4	Y3	Y2	Y3	Y2	Y1	Y4	Y4	Y1	Y3	Y2	Y2	Y3	Y1	Y4
AII	AII	AII	AII	CII	CII	CII	CII	PII	PII	PII	PII	LII	LII	LII	LII
Y1	Y4	Y3	Y2	Y3	Y2	Y1	Y4	Y4	Y1	Y3	Y2	Y2	Y3	Y1	Y4
AII	AII	AII	AII	CII	CII	CII	CII	PII	PII	PII	PII	LII	LII	LII	LII
Z1	Z4	Z3	Z2	Z3	Z2	Z4	Z1	Z4	Z1	Z3	Z2	Z2	Z3	Z1	Z4
AII	AII	AII	AII	CII	CII	CII	CII	PII	PII	PII	PII	LII	LII	LII	LII
Z1	Z4	Z3	Z2	Z3	Z2	Z4	Z1	Z4	Z1	Z3	Z2	Z2	Z3	Z1	Z4
S2	S2	S2	S2	S2	S2	S2	S2	S2	S2	S2	S2	S2	S2	S2	S2
AIII	AIII	AIII	AIII	CIII	CIII	CIII	CIII	PIII	PIII	PIII	PIII	LIII	LIII	LIII	LIII
Y1	Y4	Y3	Y2	Y3	Y2	Y4	Y1	Y4	Y1	Y2	Y3	Y2	Y3	Y4	Y1
AIII	AIII	AIII	AIII	CIII	CIII	CIII	CIII	PIII	PIII	PIII	PIII	LIII	LIII	LIII	LIII
Y4	Y1	Y2	Y3	Y3	Y2	Y4	Y1	Y4	Y1	Y2	Y3	Y2	Y3	Y4	Y1
AIII	AIII	AIII	AIII	CIII	CIII	CIII	CIII	PIII	PIII	PIII	PIII	LIII	LIII	LIII	LIII
Z4	Z1	Z2	Z3	Z3	Z2	Z4	Z1	IZ4	Z1	Z2	Z3	Z2	Z3	Z4	Z1
AIII	AIII	AIII	AIII	CIII	CIII	CIII	CIII	PIII	PIII	PIII	PIII	LIII	LIII	LIII	LIII
Z4	Z1	Z2	Z3	Z3	Z2	Z4	Z1	Z4	Z1	Z2	Z3	Z2	Z3	Z4	Z1
AI	AI	AI	AI	CI	CI	CI	CI	PI	PI	PI	PI	LI	LI	LI	LI
Z1	Z4	Z3	Z2	Z2	Z3	Z4	Z1	Z1	Z4	Z3	Z2	Z3	Z2	Z1	Z4
AI	AI	AI	AI	CI	CI	CI	CI	PI	PI	PI	PI	LI	LI	LI	LI
Z1	Z4	Z3	Z2	Z2	Z3	Z4	Z1	Z1	Z4	Z3	Z2	Z3	Z2	Z1	Z4
AI	AI	AI	AI	CI	CI	CI	CI	PI	PI	PI	PI	LI	LI	LI	LI
Y1	Y4	Y3	Y2	Y2	Y3	Y4	Y1	Y1	Y4	Y3	Y2	Y3	Y2	Y1	Y4
AI	AI	AI	AI	CI	CI	CI	CI	PI	PI	PI	PI	LI	LI	LI	LI
Y1	Y4	Y3	Y2	Y2	Y3	Y4	Y1	Y1	Y4	Y3	Y2	Y3	Y2	Y1	Y4

of variance were discussed in Chapter 5. Additional discussion of these rules and their use may be found in Federer (1975, 1977, and 1984). As may be surmised, the partitioning of the degrees of freedom for an experiment as complex as this one will be a tedious and thought provoking task.

Table 6.1. (*Continued*)

Replicate 3 or row 3

Old cane land   Old pineapple land

CIII	CIII	CIII	CIII	AIII	AIII	AIII	AIII	LIII	LIII	LIII	LIII	PIII	PIII	PIII	PIII
Z4	Z1	Z2	Z3	Z3	Z2	Z1	Z4	Z4	Z1	Z3	Z2	Z3	Z2	Z1	Z4
CIII	CIII	CIII	CIII	AIII	AIII	AIII	AIII	LIII	LIII	LIII	LIII	PIII	PIII	PIII	PIII
Z4	Z1	Z2	Z3	Z3	Z2	Z1	Z4	Z4	Z1	Z3	Z2	Z3	Z2	Z1	Z4
CIII	CIII	CIII	CIII	AIII	AIII	AIII	AIII	LIII	LIII	LIII	LIII	PIII	PIII	PIII	PIII
Y4	Y1	Y2	Y3	Y3	Y2	Y1	Y4	Y4	Y1	Y3	Y2	Y3	Y2	Y1	Y4
CIII	CIII	CIII	CIII	AIII	AIII	AIII	AIII	LIII	LIII	LIII	LIII	PIII	PIII	PIII	PIII
Y4	Y1	Y2	Y3	Y3	Y2	Y1	Y4	Y4	Y1	Y3	Y2	Y3	Y2	Y1	Y4
CI	CI	CI	CI	AI	AI	AI	AI	LI	LI	LI	LI	PI	PI	PI	PI
Y4	Y1	Y3	Y2	Y2	Y3	Y4	Y1	Y1	Y4	Y2	Y3	Y2	Y3	Y4	Y1
CI	CI	CI	CI	AI	AI	AI	AI	LI	LI	LI	LI	PI	PI	PI	PI
Y4	Y1	Y3	Y2	Y2	Y3	Y4	Y1	Y1	Y4	Y2	Y3	Y2	Y3	Y4	Y1
CI	CI	CI	CI	AI	AI	AI	AI	LI	LI	LI	LI	PI	PI	PI	PI
Z4	Z1	Z3	Z2	Z2	Z3	Z4	Z1	Z1	Z4	Z2	Z3	Z2	Z3	Z4	Z1
CI	CI	CI	CI	AI	AI	AI	AI	LI	LI	LI	LI	PI	PI	PI	PI
Z4	Z1	Z3	Z2	Z2	Z3	Z4	Z1	Z1	Z4	Z2	Z3	Z2	Z3	Z4	Z1
CII	CII	CII	CII	AII	AII	AII	AII	LII	LII	LII	LII	PII	PII	PII	PII
Y1	Y4	Y2	Y3	Y3	Y2	Y1	Y4	Y4	Y1	Y3	Y2	Y3	Y2	Y1	Y4
CII	CII	CII	CII	AII	AII	AII	AII	LII	LII	LII	LII	PII	PII	PII	PII
Y1	Y4	Y2	Y3	Y3	Y2	Y1	Y4	Y4	Y1	Y3	Y2	Y3	Y2	Y1	Y4
CII	CII	CII	CII	AII	AII	AII	AII	LII	LII	LII	LII	PII	PII	PII	PII
Z1	Z4	Z2	Z3	Z3	Z2	Z1	Z4	Z4	Z1	Z3	Z2	Z3	Z2	Z1	Z4
CII	CII	CII	CII	AII	AII	AII	AII	LII	LII	LII	LII	PII	PII	PII	PII
Z1	Z4	Z2	Z3	Z3	Z2	Z1	Z4	Z4	Z1	Z3	Z2	Z3	Z2	Z1	Z4

6.3. PRELIMINARY ANALYSES FOR THE EXPERIMENT

As a first step in unraveling the total degrees of freedom into its component parts, consider only the four row by four column arrangement with the four treatments C, A, P, and L as given in Figure 6.1. This four-row by four-column

Table 6.1. (*Continued*)

Replicate 4 or row 4

Old cane land   Old pineapple land

AII	AII	AII	AII	CII	CII	CII	CII	PII	PII	PII	PII	LII	LII	LII	LII
Z4	Z1	Z2	Z3	Z2	Z3	Z4	Z1	Z1	Z4	Z3	Z2	Z3	Z2	Z4	Z1
AII	AII	AII	AII	CII	CII	CII	CII	PII	PII	PII	PII	LII	LII	LII	LII
Z4	Z1	Z2	Z3	Z2	Z3	Z4	Z1	Z1	Z4	Z3	Z2	Z3	Z2	Z4	Z1
AII	AII	AII	AII	CII	CII	CII	CII	PII	PII	PII	PII	LII	LII	LII	LII
Y4	Y1	Y2	Y3	Y2	Y3	Y4	Y1	Y1	Y4	Y3	Y2	Y3	Y2	Y4	Y1
AII	AII	AII	AII	CII	CII	CII	CII	PII	PII	PII	PII	LII	LII	LII	LII
Y4	Y1	Y2	Y3	Y2	Y3	Y4	Y1	Y1	Y4	Y3	Y2	Y3	Y2	Y4	Y1
AI	AI	AI	AI	CI	CI	CI	CI	PI	PI	PI	PI	LI	LI	LI	LI
Y1	Y4	Y3	Y2	Y2	Y3	Y1	Y4	Y1	Y4	Y2	Y3	Y2	Y3	Y1	Y4
AI	AI	AI	AI	CI	CI	CI	CI	PI	PI	PI	PI	LI	LI	LI	LI
Y1	Y4	Y3	Y2	Y2	Y3	Y1	Y4	Y1	Y4	Y2	Y3	Y2	Y3	Y1	Y4
AI	AI	AI	AI	CI	CI	CI	CI	PI	PI	PI	PI	LI	LI	LI	LI
Z1	Z4	Z3	Z2	Z2	Z3	Z1	Z4	Z1	Z4	Z2	Z3	Z2	Z3	Z1	Z4
AI	AI	AI	AI	CI	CI	CI	CI	PI	PI	PI	PI	LI	LI	LI	LI
Z1	Z4	Z3	Z2	Z2	Z3	Z1	Z4	Z1	Z4	Z2	Z3	Z2	Z3	Z1	Z4
AIII	AIII	AIII	AIII	CIII	CIII	CIII	CIII	PIII	PIII	PIII	PIII	LIII	LIII	LIII	LIII
Y1	Y4	Y2	Y3	Y3	Y2	Y1	Y4	Y4	Y1	Y3	Y2	Y3	Y2	Y1	Y4
AIII	AIII	AIII	AIII	CIII	CIII	CIII	CIII	PIII	PIII	PIII	PIII	LIII	LIII	LIII	LIII
Y1	Y4	Y2	Y3	Y3	Y2	Y1	Y4	Y4	Y1	Y3	Y2	Y3	Y2	Y1	Y4
AIII	AIII	AIII	AIII	CIII	CIII	CIII	CIII	PIII	PIII	PIII	PIII	LIII	LIII	LIII	LIII
Z1	Z4	Z2	Z3	Z3	Z2	Z1	Z4	Z4	Z1	Z3	Z2	Z3	Z2	Z1	Z4
AIII	AIII	AIII	AIII	CIII	CIII	CIII	CIII	PIII	PIII	PIII	PIII	LIII	LIII	LIII	LIII
Z1	Z4	Z2	Z1	Z3	Z2	Z1	Z4	Z4	Z1	Z3	Z2	Z3	Z2	Z1	Z4

arrangement is not a Latin square design even if it appears so. There are several ways of partitioning the 16 degrees of freedom into component parts. Two of these are given in Tables 6.2 and 6.3 and a third in Table 6.4. The contrasts of $C + A$ versus $P + L$ and columns $1 + 2$ versus columns $3 + 4$ are completely confounded.

Table 6.2. Partitioning of Degrees of Freedom in an Analysis of Variance.

Source of variation	Degrees of freedom
Total	16
Correction for the mean	1
Soils (Columns 1 + 2 versus 3 + 4 = C + A versus P + L) = S	1
Columns within soils = error S	2
Rows = R	3
Treatments on cane land (C versus A) = CA	1
Treatments on pineapple land (P versus L) = PL	1
Residual = error CA and PL	7

Table 6.3. Partitioning of Degrees of Freedom in an Analysis of Variance.

Source of variation	Degrees of freedom
Total	16
Correction for the mean	1
Soils (Columns 1 + 2 versus 3 + 4 = C + A versus P + L) = S	1
Columns within soils = error S	2
Treatments on cane land (C versus A) = CA	1
Treatments on pineapple land (P versus L) = PL	1
Between duplicates for C in columns 1 and 2	2
Between duplicates for A in columns 1 and 2	2
Between duplicates for P in columns 3 and 4	2
Between duplicates for L in columns 3 and 4	2
CA × columns on sugar cane land	1
PL × columns on pineapple land	1

Table 6.4. Partitioning the 16 Degrees of Freedom for the Two Four Rows by Two Columns Experiment Designs.

Source of variation	Degrees of freedom
Total	16
Correction for the mean	1
Soils = columns 1 + 2 versus 3 + 4 = C + A versus P + L = S	1
Error S = columns within S	2
C versus A = CA	1
Rows within sugar cane land = R/C	3
Error CA	2
P versus L = PL	1
Rows within pineapple land = R/P	3
Error PL	2

Owing to the nature of the layout of these 16 plots, a split plot arrangement is obtained with soils being the whole plot treatments and C and A are split plot treatments within sugar cane land and P and L are split plot treatments within pineapple land. Hence, there are two error terms, error S and error CA and PL. The seven residual degrees of freedom are composed of CA × rows within sugar cane land with three degrees of freedom (error CA), PL × rows within pineapple land with three degrees of freedom (error PL), and one degree of freedom owing to the fact that there is complete confounding of C + A versus P + L and columns 1 + 2 versus columns 3 + 4. This single degree of freedom represents the contrast between the four columns on sugar cane land and the four columns on pineapple land.

A second partitioning of degrees of freedom is presented in Table 6.3. Note that other partitions, such as 16 single degrees of freedom contrasts, may be reasonable. Although a rationale can be formed for each of the analysis, the most appropriate breakdown of these 16 degrees of freedom is to consider that a four row by two column design with two treatments was laid out on each of the two types of land or soil. Then a combined analysis is made for the two experiment designs. Such a partitioning is given in Table 6.4. We shall use this partitioning for the combined analysis.

Since one degree of freedom for columns is subtracted from the CA × R/C and from PL × R/P degrees of freedom, there are only two degrees of freedom left for the error terms for CA and PL. For the contrast of sugar cane soil versus pineapple soil, S, the comparison of columns within S is used for Error S.

Ignoring the factors preplanting fumigation (I, II, III), postplanting fumigation (Y, Z), and postplanting fertilization (S, U), an analysis of variance partitioning of the degrees of freedom for the remaining factors (Figure 6.5) is given in Table 6.5. There are 15 different error terms in this partitioning of the 64 degrees of freedom.

From Figure 6.5, it can be seen that the replicates for the contrast of major elements versus minor elements M are the columns, cols, and not the rows R. For mi there are four columns. For treatment 1 versus 4, ma, there are four columns. The same holds true for the contrast of treatment 2 versus treatment 3, mi.

Ignoring the treatment factors in Table 6.5, the factors preplanting fumigation (I, II, III), postplanting fumigation (Y, Z), and postplanting fertilization (S, U) are in a standard split split plot experiment design arrangement where the rows form the replicates. An analysis of variance for this part of the experiment is given in Table 6.6. As for a standard split split plot experiment design, there are three different error terms in this partitioning. There are a total of 48 observations in this breakdown of the degrees of freedom. The rows form the replicates for these three sets of treatments.

Other partitions of parts of the experiment may be useful before proceeding to the combined analysis of variance for partitioning the degrees of freedom for the 768 observations. For example, it may be useful to consider an analysis for the original sixteen plots together with the three preplanting fumigation treatments and ignoring all other factors. This consideration involves a form of a split block experiment design.

Table 6.5. Partitioning of the Degrees of Freedom in an Analysis of Variance Table for Soils, Treatments 1 to 4, Rows, Major versus Minor Elements, Major Elements, and Minor Elements.

Source of variation	Degrees of freedom	Subtotal
Total	64	
Correction for the mean	1	
Soils (Columns $1 + 2$ versus $3 + 4 = C + A$ versus $P + L$) = S	1	
Columns within soils = error S	2	
Rows on sugar cane land = R/C	3	
Treatments on cane land (C versus A) = CA	1	
Error CA	2	
Rows on pineapple land	3	
Treatments on pineapple land (P versus L) = PL	1	
Error PL	2	16
Major versus minor = $1 + 4$ versus $2 + 3 = M$	1	
Error M = M × cols	3	
M × CA	1	
Error M × CA = M × CA × R	3	
M × PL	1	
Error M × PL = M × PL × R	3	
M × S	1	
Error M × S = M × S × cols	3	32
Major = ma = 1 versus 4	1	
Error ma = ma × cols	3	
ma × CA	1	
Error ma × CA = ma × CA × R	3	
ma × PL	1	
Error ma × PL = ma × PL × R	3	
ma × S	1	
Error ma × S = ma × S × cols	3	
Minor = mi = 2 versus 3	1	
Error mi = mi × cols	3	
mi × CA	1	
Error mi × R	3	
mi × PL	1	
Error mi × PL = mi × PL × R	3	
mi × S	1	
Error mi × S = mi × S × cols	3	64

A COMBINED ANALYSIS OF VARIANCE PARTITIONING

Table 6.6. Partitioning the Degrees of Freedom in an Analysis of Variance for the Treatment Factors Preplanting Fumigation (I, II, III), Postplanting Fumigation (Y, Z), and Postplanting Fertilization (S and U).

Source of variation	Degrees of freedom	Subtotal
Total	48	
Correction for the mean	1	
Rows = replicates = R	3	
Preplanting fumigation = F	2	
Error F = F × R	6	12
Postplanting fumigation = Y versus Z = PF	1	
F × PF	2	
Error PF = PF × R within F	9	24
Postplanting fertilization = U versus S = U	1	
U × F	2	
U × PF	1	
U × F × PF	2	
Error U = U × R within F and PF	18	48

The rules and procedures used above to obtain statistical analyses have been explained in detail by Federer (1975, 1977) and in Chapter 5.

6.4. A COMBINED ANALYSIS OF VARIANCE PARTITIONING OF THE DEGREES OF FREEDOM

Note that the factors in Table 6.6 are in a split block arrangement with those in Table 6.5. There are a total of 12 × 64 = 768 observations. A combined analysis of variance partitioning of the degrees of freedom is presented in Table 6.7. We first enter the analysis of variance table in Table 6.5 with the partitioning of the 64 degrees of freedom. Then, the partitioning of the degrees of freedom in Table 6.6 is added. To this point 108 degrees of freedom have been taken into account. There are 64 coming from Table 6.5 and 44 from Table 6. Four of the degrees of freedom in Table 6.6 have been included in Table 6.5. These are three degrees of freedom for rows and one degree of freedom for the correction for the mean. Rows are taken to be the replicates for factors CA, PL, F, PF, and U. Columns form the replicates for the other factors.

Note that the factor preplanting fumigation was first combined with the factors in Table 6.5. Then the factor postplanting fumigation was added, and finally the factor postplanting fertilization was added to complete the partitioning. There are

Table 6.7. A Partitioning of the 768 Degrees of Freedom for the Experiment.

Source of variation	Degree of freedom	Obser.	Error term	Line
Total	768			1
Correction for the mean	1			2
Soils (Columns 1 + 2 versus 3 + 4 = C + A versus P + L) = S	1			3
Columns within soils = error S	2			4
Rows = R/sugar cane = R/C	3			5
Treatments on cane land (C versus A) = CA	1			6
Error CA	2		E1	7
Treatments on pineapple land (P versus L) = PL	1			8
Rows on pineapple land = R/P	3			9
Error PL	2	16	E2	10
Major versus minor = 1 + 4 versus 2 + 3 = M	1			11
Error M = M × cols	3		E3	12
M × CA	1			13
Error M × CA = M × CA × R	3		E4	14
M × PL	1			15
Error M × PL = M × PL × R	3		E5	16
M × S	1			17
Error M × S = M × S × cols	3	32	E6	18
Major = ma = 1 versus 4	1			19
Error ma = ma × cols	3		E7	20
ma × CA	1			21
Error ma × CA = ma × CA × R	3		E8	22
ma × PL	1			23
Error ma × PL = ma × PL × R	3		E9	24
ma × S	1			25
Error ma × S = ma × S × cols	3		E10	26
Minor = mi = 2 versus 3	1			27
Error mi = mi × cols	3		E11	28
mi × CA	1			29
Error mi × R	3		E12	30
mi × PL	1			31
Error mi × PL = mi × PL × R	3		E13	32
mi × S	1			33
Error mi × S = mi × S × cols	3	64	E14	34
Preplanting fumigation = F	2			35
Error F = F × R	6	72	E15	36
Postplanting fumigation = Y versus Z = PF	1			37

Table 6.7. (*Continued*)

Source of variation	Degree of freedom	Obser.	Error term	Line
F × PF	2			38
Error PF = PF × R within F	9	84	E16	39
Postplanting fertilization = U versus S = U	1			40
U × F	2			41
U × PF	1			42
U × F × PF	2			43
Error U = U × R within F and PF	18	108	E17	44
F × S	2			45
Error FS = F × S × R	6		E18	46
F × CA	2			47
Error F × CA = F × CA × R/C	6		E19	48
F × PL	2			49
Error F × PL = F × PL × R/P	6		E20	50
F × M	2			51
Error F × M = F × M × R	6		E21	52
F × M × CA	2			53
Error F × M × CA = F × M × CA × R	6		E22	54
F × M × PL	2			55
Error F × M × PL = F × M × PL × R	6		E23	56
F × M × S	2			57
Error F × M × S = F × M × S × R	6		E24	58
F × ma	2			59
Error F × ma = F × ma × R	6		E25	60
F × ma × CA	2			61
Error F × ma × CA = F × ma × CA × R	6		E26	62
F × ma × PL	2			63
Error = F × ma × PL × R	6		E27	64
F × ma × S	2			65
Error = F × ma × S × R	6		E28	66
F × mi	2			67
Error = F × mi × R	6		E29	68
F × mi × CA	2			69
Error = F × mi × CA × R	6		E30	70
F × mi × PL	2			71
Error = F × mi × PL × R	6		E31	72
F × mi × S	2			73
Error = F × mi × S × R	6	228	E32	74
PF × S	1			75
PF × F × S	2			76
Error = PF × S × R within F	9		E33	77

Table 6.7. (*Continued*)

Source of variation	Degree of freedom	Obser.	Error term	Line
PF × CA	1			78
PF × F × CA	2			79
Error = PF × CA × R within F	9		E34	80
PF × PL	1			81
PF × F × PL	2			82
Error = PF × PL × R within F	9		E35	83
PF × M	1			84
PF × F × M	2			85
Error = PF × M × R within F	9		E36	86
PF × M × S	1			87
PF × F × M × S	2			88
Error = PF × M × S within F	9		E37	89
PF × M × CA	1			90
PF × F × M × CA	2			91
Error = PF × M × CA × R within F	9		E38	92
PF × M × PL	1			93
PF × F × M × PL	2			94
Error = PF × M × PL × R within F	9		E39	95
PF × ma	1			96
PF × F × ma	2			97
Error = PF × ma × R within F	9		E40	98
PF × ma × S	1			99
PF × F × ma × S	2			100
Error = PF × ma × R within F	9		E41	101
PF × ma × CA	1			102
PF × F × ma × CA	2			103
Error = PF × ma × CA × R within F	9		E42	104
PF × ma × PL	1			105
PF × F × ma × PL	2			106
Error = PF × ma × PL × R within F	9		E43	107
PF × mi	1			108
PF × F × mi	2			109
Error = PF × mi × R within F	9		E44	110
PF × mi × S	1			111
PF × F × mi × S	2			112
Error = PF × mi × R within F	9		E45	113
PF × mi × CA	1			114
PF × F × mi × CA	2			115
Error = PF × mi × CA × R within F	9		E46	116
PF × mi × PL	1			117
PF × F × mi × PL	2			118
Error = PF × mi × CA × R within F	9	408	E47	119
U × S	1			120
U × F × S	2			121

A COMBINED ANALYSIS OF VARIANCE PARTITIONING

Table 6.7. (*Continued*)

Source of variation	Degree of freedom	Obser.	Error term	Line
U × PF × S	1			122
U × F × PF × S	2			123
Error = U × S × R within F and PF	18		E48	124
U × CA	1			125
U × F × CA	2			126
U × PF × CA	1			127
U × F × PF × CA	2			128
Error = U × CA × R within F and PF	18		E49	129
U × PL	1			130
U × F × PL	2			131
U × PF × PL	1			132
U × F × PF × PL	2			133
Error = U × PL × R	18		E50	134
U × M	1			135
U × F × M	2			136
U × PF × M	1			137
U × F × PF × M	2			138
Error = U × M × R within F and PF	18		E51	139
U × M × S	1			140
U × F × M × S	2			141
U × PF × M × S	1			142
U × F × PF × M × S	2			143
Error = U × M × S × R within F and PF	18		E52	144
U × M × CA	1			145
U × F × M × CA	2			146
U × PF × M × CA	1			147
U × F × PF × M × CA	2			148
Error = U × M × CA × R within F and PF	18		E53	149
U × M × PL	1			150
U × F × M × PL	2			151
U × PF × M × PL	1			152
U × F × PF × M × PL	2			153
Error = U × M × PL × R within F and PF	18		E54	154
U × ma	1			155
U × F × ma	2			156
U × PF × ma	1			157
U × F × PF × ma	2			158
Error = U × ma × R within F and PF	18		E55	159
U × ma × S	1			160
U × F × ma × S	2			161
U × PF × ma × S	1			162

Table 6.7. (*Continued*)

Source of variation	Degree of freedom	Obser.	Error term	Line
U × F × PF × ma × S	2			163
Error = U × ma × S × R within F and PF	18		E56	164
U × ma × CA	1			165
U × F × ma × CA	2			166
U × PF × CA	1			167
U × F × PF × CA	2			168
Error = U × ma × CA × R within F and PF	18		E57	169
U × ma × PL	1			170
U × F × ma × PL	2			171
U × PF × ma × PL	1			172
U × F × PF × ma × PL	2			173
Error = U × ma × PL × R within F and PF	18		E58	174
U × mi	1			175
U × F × mi	2			176
U × PF × mi	1			177
U × F × PF × mi	2			178
Error = U × mi × R within F and PF	18		E59	179
U × mi × S	1			180
U × F × mi × S	2			181
U × PF × mi × S	1			182
U × F × PF × mi × S	2			183
Error = U × mi × S × R within F and PF	18		E60	184
U × mi × CA	1			185
U × F × mi × CA	2			186
U × PF × mi × CA	1			187
U × F × PF × mi × CA	2			188
Error = U × mi × CA × R	18		E61	189
U × mi × PL	1			190
U × F × mi × PL	2			191
U × PF × mi × PL	1			192
U × F × PF × mi × PL	2			193
Error = U × mi × PL × R within F and PF	18	768	E62	194

62 error terms in the above partitioning of the 768 degrees of freedom. Note that 65 lines are easily added to this table by using the contrast for F of none versus the mean of the 400 and 800 pound applications and the contrast of 400 versus 800 and all interactions involving these contrasts. This would have produced an analysis of variance table with $194 + 65 = 259$ lines.

6.5. SOME COMMENTS

Since such analyses, as presented above, involve a considerable amount of creative thinking and effort on the part of the statistical analyst, joint authorship on publication of results from the experiment should be the rule. Most experimenters would be unable to produce analyses such as those described in this and the previous three chapters. This is not standard textbook material. Hence, the aid of a statistician is crucial in summarizing the results from such experiments.

6.6. PROBLEMS

Problem 6.1. Ignore all row treatment applications and obtain a partitioning of the degrees of freedom in the analysis of variance using the results in Table 6.2. Describe the type of design for this analysis.

Problem 6.2. Ignore all column treatment applications and obtain a partitioning of the degrees of freedom in the analysis of variance using the results in Table 6.3. Describe the design of the experiment for this analysis.

Problem 6.3. Using the analysis of variance given in Table 6.2, obtain a partitioning of the degrees of freedom in an analysis of variance. How many lines are there in this analysis of variance? How many error terms are there?

6.7. REFERENCES

Federer, W. T. (1975). The misunderstood split plot. In *Applied Statistics* (Editor: R. P. Gupta), North Holland Publishing Company, Amsterdam, pp. 9–38.

Federer, W. T. (1977). Sampling, blocking, and model considerations for split plot and split block designs. Biometrical Journal 19(3):181–200.

Federer, W. T. (1984). Principles of statistical design with special reference to experiment and treatment design. In *Statistics: An Appraisal*, David, H. A. and H. T. David, Editors, The Iowa State University Press, Ames, Iowa, pp. 77–104.

Federer, W. T. and C. A. Farden. (1955). Analysis of variance set-up for Joint Project 69. BU-67-M in the Technical Report Series of the Biometrics Unit, *Cornell University*, Ithaca, New York 14853.

APPENDIX 6.1. FIGURE 6.1 TO FIGURE 6.6

Old sugarcane land		Roadway	Old pineapple land	
C	A		L	P
A	C		P	L
C	A		L	P
A	C		P	L

Figure 6.1. Original 16 experimental units.

Preplanting fumigation	Old sugarcane land		Roadway	Old pineapple land	
I II III	C	A		L	P
II I III	A	C		P	L
III I II	C	A		L	P
II I III	A	C		P	L

Figure 6.2. Adding preplanting fumigation as a split block treatment.

Preplanting fumigation	Postplanting fumigation	Old sugarcane land	Roadway		Old pineapple land	
I	Y	**C**	**A**		**P**	**L**
	Z					
II	Z					
	Y					
III	Y					
	Z					
II	Y	**A**	**C**		**P**	**L**
	Z					
I	Y					
	Z					
III	Z					
	Y					
III	Z	**C**	**A**		**L**	**P**
	Y					
I	Y					
	Z					
II	Y					
	Z					
II	Z	**A**	**C**		**P**	**L**
	Y					
I	Y					
	Z					
III	Y					
	Z					

Figure 6.3. Adding postplanting fumigation as a split plot to preplanting fumigation.

Preplanting fumigation	Postplanting fumigation	Fertilizer	Old sugarcane land	Roadway	Old pineapple land	
I	Y	U S	C	A	L	P
	Z	U S				
II	Z	S U				
	Y	U S				
III	Y	U S				
	Z	S U				
II	Y	S U	A	C	P	L
	Z	U S				
III	Y	U S				
	Z	S U				
I	Z	S U				
	Y	U S				
III	Z	S U	C	A	L	P
	Y	S U				
I	Y	U S				
	Z	S U				
II	Y	U S				
	Z	U S				
II	Z	S U	A	C	P	L
	Y	U S				
I	Y	U S				
	Z	S U				
III	Y	S U				
	Z	U S				

Figure 6.4. Adding fertilizer (split split plot) treatments to the layout in Figure 6.3.

	Pre-Plant fumig.	Old cane land major	Old cane land minor elem.	major	Roadway	Old pineapple land major	Old pineapple land minor elem.	major	Post plant fumig.
I		/	/			/	/		Y Z
II									Z Y
III		/	/			/	/		Y Z
II		/	/			/	/		Y Z
III									Z Y
I		/	/			/	/		Z Y
III		/	/			/	/		Z Y
I									Y Z
II		/	/			/	/		Y Z
II		/	/			/	/		Z Y
I									Y Z
III		/	/			/	/		Y Z

Figure 6.5. Design of experiment in columns illustrating the layout of the experiment with major and minor elements added and S and U omitted. Treatments 1 and 4 appear as major and 2 and 3 appear as minor treatments.

WORLD RECORDS FOR THE LARGEST ANALYSIS

Pre-plant fumig.	Post-plant fumig.	Post-plant fert.	Old cane land			Road-way	Old pineapple land plant		
			Major	Minor	element Major		Major	Minor	element Major
I	Y	U S	/	/			/	/	
	Z	S U S							
II	Z	S U U S	/	/			/	/	
	Y								
III	Y	U S S U	/	/			/	/	
	Z								
II	Y	S U S U	/	/			/	/	
	Z								
III	Y	U S S U	/	/			/	/	
	Z								
I	Z	S U U S	/	/			/	/	
	Y								
III	Z	S U S U	/	/			/	/	
	Y								
I	Y	U S S U	/	/			/	/	
	Z								
II	Y	U S U S	/	/			/	/	
	Z								
II	Z	S U U S	/	/			/	/	
	Y								
I	Y	U S S U	/	/			/	/	
	Z								
III	Y	S U U S	/	/			/	/	
	Z								

Figure 6.6. Layout of the experiment with all treatments added except C, A, P, and L. Treatments 1 and 4 appear randomly as major, and 2 and 3 appear as minor treatments.

CHAPTER 7

Augmented Split Plot Experiment Design

7.1. INTRODUCTION

A number of different augmented experiment designs have been described in the literature. Starting with the paper by Federer (1956), augmented experiment designs were created for screening large number of genotypes or other treatments such as herbicides, fungicides, and others. An augmented experiment design is obtained by starting with any standard experiment design plan and then increasing the block size, the number of rows and/or columns, and others, in order to include n new or augmented treatments, which are usually included only once in an experiment. The large number of new treatments and scarcity of material preclude the new treatments from being replicated. Augmented complete and incomplete block experiment designs were described by Federer (1961). Augmented Latin square designs were presented by Federer and Raghavarao (1975), Federer et al. (1975), and Federer (1998). A presentation of augmented designs is given in Chapter 10 of the Federer (1993) book on intercropping experiments. The class of augmented lattice square experiment designs has been presented by Federer (2002). SAS codes for analyzing data from augmented experiment designs have been described by Wolfinger et al. (1997), Federer (2003), and Federer and Wolfinger (2003). Federer (2005) has described a class of augmented split block designs and has described analyses for these designs and presented a SAS code for the associated analyses.

The contents of this chapter have been given by Federer and Arguillas (2005a,b), where they have shown how to construct and analyze data from the class of *augmented split plot experiment designs* (ASPEDs) and to present some variations for this class of designs. In the following section, Section 7.2, a

---

Variations on Split Plot and Split Block Experiment Designs, by Walter T. Federer and Freedom King
Copyright © 2007 John Wiley & Sons, Inc.

description of an augmented randomized experiment design with split plots is given. The augmented and related check treatments, genotypes, form the whole plots. The statistical analysis is presented and is illustrated with a numerical example.

In Section 7.3, the genotype treatments form the split plot treatments in an ASPED. The statistical analysis is discussed and a numerical example is presented to illustrate the procedure. In Section 7.4, the design of Section 7.3 with split split plot treatments is discussed and a statistical analysis illustrated with a numerical example is presented. Other variations are described in Section 7.5. Computer codes, data, and output for each of the three numerical examples are given in the three appendices.

7.2. AUGMENTED GENOTYPES AS THE WHOLE PLOTS

ASPEDs may be used for any type of treatment for screening purposes. We use genotype as the treatment and do not infer that an ASPED is useful only for genotypes as treatments. To illustrate an experiment with augmented genotypes as whole plots, suppose n new genotypes are to be screened for their performance using varying levels of density, irrigation, insecticides, herbicides, date of planting, or other variables that are used in the growing of a cultivar. These genotypes may have been screened in a previous cycle or cycles and have been reduced in number. At this stage of selection, it may be desirable to consider all genotypes in an experiment as fixed effects. Consider an augmented randomized complete block experiment design with r replicates, c standards or checks, and n new genotypes with n/r new genotypes per replicate. Each of the new genotypes appears only once in the experiment.

Using a form of parsimonious experiment design as described by Federer and Scully (1993) and Federer (1993), a whole plot experimental unit (wpeu) would be planted in such a manner that the split plot treatment has decreasing values throughout the wpeu. For example, suppose that the wpeu is 30 feet long. Then the split plot treatment could be applied in decreasing amounts throughout the wpeu. Or, one could partition the wpeu into five, say, split plot experimental units each measuring six feet in length with either decreasing amounts or by randomizing the levels of the split plot treatment. Certain treatments, say date of planting, density, irrigation, herbicide, and fertilizer, lend themselves better to a continuous decrease. For such cases, there should be no natural gradients within the wpeus. If there is only random variation within a whole plot experimental unit, a systematic application of the split plot treatments from one end of the experimental unit to the other would not affect the statistical analysis and a regression response equation may suffice. It would not be advisable to have a high level of fertilizer adjacent to a low level unless the plots are reasonably far apart—about the height of the plant—to avoid inter-plot competition.

Without partitioning the genotype degrees of freedom into check and new, a standard split plot linear model holds, as can be seen from the following table.

An analysis of variance table for the above-described experiment may be of the form:

Source of variation	Degrees of freedom
Total	$f(rc+n)$
Correction for mean	1
Replicates, R	$r-1$
Genotypes, whole plot treatments G	$c+n-1$
Checks C	$c-1$
New N	$n-1$
C versus N	1
$G \times R =$ checks $\times R = C \times R$	$(c-1)(r-1)$
Date of planting D	$d-1$
$D \times G$	$(d-1)(c+n-1)$
$D \times C$	$(d-1)(c-1)$
$D \times N$	$(d-1)(n-1)$
$D \times C$ versus N	$d-1$
$D \times R$ within G	$c(d-1)(r-1)$

More information is obtained on the date of planting D and on the date of planting by genotype interaction $D \times G$ than on genotypes in a design such as this.

Example 7.1. A numerical example is presented to illustrate an analysis for an ASPED with genotypes as the whole plot treatments in an augmented randomized complete block experiment design and with fertilizer as the split plot treatments. For the genotypes, let the number of check genotypes be $c=4$ (numbered 25–28), the number of new genotypes be $n=24$ (numbered 1–24), the number of fertilizers be $f=3$ (F1, F2, and F3), and the number of replicates or blocks be $r=4$. Owing to the fact that the SAS PROC GLM procedure (SAS Institute, 1999–2001) sets the highest numbered effect equal to zero, it is a good idea to use the highest number for a check genotype as all the estimated effects have the highest numbered effect subtracted from all other effects. The effects obtained in the SAS output are differences of an effect minus the highest numbered effect. The standard error given is a standard error of a difference between two effects and not a standard error of an effect or of a mean, as indicated on the output.

Artificial data for an ASPED with four check genotypes, 24 new genotypes, three fertilizer levels, and with four replicates are presented in Table 7.1.

The SAS code for the above data is given in Appendix 7.1. An analysis of variance table, ANOVA, for the responses in Table 7.1 is presented in Table 7.2. Type III sums of squares and mean squares are reported. Also presented in Table 7.2 is an ANOVA for checks only and for new treatments only. The sources of variation in the ANOVA tables are based on and describe the linear model utilized for

Table 7.1. Responses (Data) for an ASPED with $f = 3$, $c = 4$, $n = 24$, and $r = 4$.

Replicate 1 F*	Genotype 25 26 27 28	1 2 3 4 5 6	Replicate 2 F*	Genotype 25 26 27 28	7 8 9 10 11 12
F1	2 1 9 9	2 3 4 5 6 7	F1	3 2 8 7	1 2 3 4 8 3
F2	4 3 9 9	5 6 7 8 8 8	F2	3 2 8 8	2 3 4 4 8 5
F3	6 5 8 7	7 5 6 4 8 7	F3	7 4 8 9	3 5 4 5 8 7

Replicate 3 F*	Genotype 25 26 27 28	13 14 15 16 17 18	Replicate 4 F*	Genotype 25 26 27 28	19 20 21 22 23 24
F1	4 2 8 7	7 5 3 7 6 5	F1	4 5 9 9	5 6 7 8 9 9
F2	6 5 7 7	7 6 5 7 8 7	F2	5 2 9 8	8 6 4 7 8 8
F3	8 7 9 9	9 8 6 9 8 8	F2	6 5 9 7	9 8 8 9 9 9

F* is fertilizer with three levels F1, F2, and F3.

the analyses. To obtain the last two ANOVAs, the SAS code is altered by putting an IF – THEN statement immediately following the INPUT statement. The SAS statement required when only checks are being considered, is

IF G < 25 THEN DELETE;

Table 7.2. ANOVA for Example 7.1, Type III Sums of Squares.

Source of variation	Degrees of freedom	Sum of squares	Mean square
Total	120	5181	—
Correction for mean	1	4575.675	—
Replicate R	3	5.750	1.917
Genotype G	27	343.483	12.722
$G \times R$	9	11.750	1.306
F	2	50.505	25.252
$F \times G$	54	77.983	1.444
$F \times R$ wn G	24	27.500	1.146
Checks only	—	—	—
Replicate R	3	5.750	1.917
Check	3	202.417	67.472
Check $\times R$	9	11.750	1.306
F	2	21.292	10.646
$F \times$ check	6	21.208	3.535
$F \times R$ within check	24	27.500	1.146
New only	—	—	—
New	23	219.986	9.565
F	2	40.444	20.222
New $\times F$	46	54.889	1.193

The SAS statement needed for only new genotypes is

IF G > 24 THEN DELETE;

The MODEL statement for new genotypes needs to be changed to

MODEL Y = F G F*G.

Some of the contrasts for new treatment effects are confounded with the replicate effects. This is the reason replicate was omitted in the above MODEL statement. The code in Appendix 7.1 for SAS PROC GLM gives solutions for the parameters in the MODEL statement and the least squares means are arranged in descending order of magnitude. The SAS PROC MIXED code results in solutions for the fixed effects in the model statement, the solutions for the random effects, the least squares means for the fixed effects in the order given in the lsmeans statement, the random effect solutions arranged in descending order, and the least squares means for fixed effects arranged in descending order. If the augmented genotypes are nearing final testing, the experimenter may wish to consider them as fixed effects rather than as random effects. In the early stages of testing, new genotypes should be considered as random effects. As they approach final testing, the selected genotypes become fixed effects with each genotype being identified by a name or number/letter designation.

Since only the check least squares means and the fertilizer by check least squares means, Table 7.3, are estimable by SAS PROC GLM, the estimated effects are presented in Table 7.4.

In the above analyses, the genotype effects have been treated as fixed effects. If the analyst wishes to consider them as random effects and uses the SAS PROC MIXED procedure, reference may be made to the codes given in Appendix 7.1, by Federer (1998, 2003), by Federer and Wolfinger (2003), and by Wolfinger et al. (1997). If the number of new genotypes becomes large, the analyst may wish to order their responses in descending order (see above references) as is done with the computer code given for this example. Both SAS PROC GLM and PROC MIXED analyses are given by the code in Appendix 7.1.

Table 7.3. Least Squares Means for Checks and Fertilizers for Example 7.1.

Genotype	F1	F2	F3	Genotype
25	3.25	4.50	6.75	4.83
26	2.50	3.00	5.25	3.58
27	8.50	8.25	8.50	8.42
28	8.00	8.00	8.00	8.00
Fertilizer	5.56	5.94	7.06	6.18

Table 7.4. Estimated Effects Using SAS GLM for Example 7.1.

Genotype	F1	F2	F3	Genotype
1	−5.00	−2.00	0.00	−1.33
2	−2.00	+1.00	0.00	−3.33
3	−2.00	+1.00	0.00	−2.33
4	+1.00	+4.00	0.00	−4.33
5	−2.00	0.00	0.00	−0.33
6	0.00	+1.00	0.00	−1.33
7	−2.00	−1.00	0.00	−5.00
8	−3.00	−2.00	0.00	−3.00
9	−1.00	0.00	0.00	−4.00
10	−1.00	−1.00	0.00	−3.00
11	+0.00	0.00	0.00	0.00
12	−4.00	−2.00	0.00	−1.00
13	−2.00	−2.00	0.00	1.33
14	−3.00	−2.00	0.00	0.33
15	−3.00	−1.00	0.00	−1.67
16	−2.00	−2.00	0.00	1.33
17	−2.00	0.00	0.00	0.33
18	−3.00	−1.00	0.00	0.33
19	−4.00	−1.00	0.00	1.00
20	−2.00	−2.00	0.00	0.00
21	−1.00	−4.00	0.00	0.00
22	−1.00	−2.00	0.00	1.00
23	−0.00	−1.00	0.00	1.00
24	0.00	−1.00	0.00	1.00
25	−3.50	−2.25	0.00	−1.08
26	−2.75	−2.25	0.00	−2.33
27	0.00	−0.25	0.00	1.08
28	0.00	0.00	0.00	0.00
Fertilizer	0.00	0.00	0.00	0.00

7.3. AUGMENTED GENOTYPES AS THE SPLIT PLOTS

The genotypes may be used as split plot treatments with whole plot treatments consisting of other factors such as tillage, fertilizer, irrigation, etc. Consider an experiment with t tillage whole plot treatments, c check or standard treatments, n new genotypes, and r replicates. The checks and new genotypes are the split plot treatments. An analysis of variance partitioning of the degrees of freedom is given in Table 7.5. There will be n/r new genotypes in each replicate if n is divisible by r. Otherwise, the number of new genotypes in a replicate will vary. A new genotype in a replicate appears in each of the t tillage whole plot treatments, but only in one of the replicates. This makes the new genotypes replicated over the t whole plot treatments.

Table 7.5. Analysis of Variance Table for t Tillage Treatments, c Check Treatments, n New Genotypes, and r Replicates with Genotypes and Checks as Split Plot Treatments.

Source of variation	Degrees of freedom
Total	$t(rc + n)$
Correction for mean	1
Replicates R	$r - 1$
Tillage T	$t - 1$
$R \times T$, Error T	$(r - 1)(t - 1)$
Genotypes G	$c + n - 1$
$G \times T$	$(t - 1)(c + n - 1)$
$\quad C \times T$	$(c - 1)(t - 1)$
$\quad N1 \times T$ within rep. 1	$(n/r - 1)(t - 1)$
$\quad N2 \times T$ within rep. 2	$(n/r - 1)(t - 1)$
$\quad \ldots$	
$\quad Nr \times T$ within rep. r	$(n/r - 1)(t - 1)$
$\quad C$ versus $N \times T$	$t - 1$
$C \times R$ within T, Error G	$t(r - 1)(c - 1)$

In the above table, Ni represents the new genotypes in each level of whole plot treatment T in replicate R_i, $i = 1, \ldots, r$. c denotes check treatments.

Example 7.2. To illustrate the analysis for an ASPED in which the genotypes are the split plot treatments, let $t = 4$ tillage whole plot treatments, $c = 3$ check genotypes numbered 20, 21, and 22, $n = 19$ new genotypes numbered 1–19, and $r = 4$ replicates. Since 19 is not divisible by 4, five new genotypes are allocated to three of the four replicates and four to the fourth replicate. There will be a total of 124 split plot experimental units, speus, and responses. There will be t replications of the particular new genotypes in each replicate as the new genotypes in a replicate appear with each of the $t = 4$ tillage whole plot treatments. Since an experiment of this type has most likely never been conducted, artificial data for this example are given in Table 7.6.

An analysis of variance for the data in Table 7.6 is given in Table 7.7. The computer code is given in Appendix 7.2. For this example, least squares means are available for tillage, genotype, and the tillage by genotype combinations. Using the SAS PROC GLM code in Appendix 7.2, the least squares means are arranged in descending order from highest to lowest. The means given in Table 7.8 are not ordered. The individual analyses of variance for checks alone and for new alone are not given, but may be obtained as described for Example 7.1. The analyses follow those for nonaugmented experiment designs.

If the new genotypes are considered to be random effects, use may be made of the PROC MIXED part of the code in Appendix 7.2. Solutions for the fixed effects in the MODEL statement, solutions for the random effects in the RANDOM statement,

Table 7.6. Responses (Data) for an ASPED with $t = 4$ Whole Plot Tillage Treatments, $r = 4$ Complete Blocks, and $c = 3$ Check Genotypes (20, 21, and 22), and $n = 19$ New Genotypes (1–19) as the Split Plot Treatments.

Replicate 1								Replicate 2							
	Genotype								Genotype						
Tillage	20	21	22	1	2	3	4	5	Tillage	20	21	22	6	7	8	9	10
1	6	4	7	8	7	6	5	9	1	8	7	6	3	3	4	4	9
2	8	5	9	9	8	5	5	8	2	9	6	7	3	2	2	5	7
3	7	6	9	5	6	7	4	6	3	7	4	5	3	3	3	6	7
4	5	4	7	3	4	5	2	4	4	6	4	7	5	4	5	7	7

Replicate 3								Replicate 4							
	Genotype								Genotype						
Tillage	20	21	22	11	12	13	14	15	Tillage	20	21	22	16	17	18	19
1	7	5	8	9	9	1	2	9	1	7	7	7	9	9	9	9
2	8	3	9	9	7	2	1	9	2	9	6	8	8	9	7	6
3	6	4	7	7	7	3	2	8	3	6	6	7	7	7	9	9
4	5	3	6	6	6	4	2	7	4	6	6	6	6	6	7	8

least squares means for the fixed effects in the MODEL statement, a descending order for the fixed effects, a descending order for the random effects, and a descending order for the fixed effect least squares means are obtained by the code.

7.4. AUGMENTED SPLIT SPLIT PLOT EXPERIMENT DESIGN

One of the many variations possible for ASPEDs is an *augmented split split plot experiment design* (ASSPED). To illustrate an ASSPED, suppose there are t treatments representing the whole plot treatments, $v = c$ checks $+ n$ new genotypes representing the split plot treatments, f split split plot treatments, and r replicates of the t whole plot treatments. That is, the ASPED in Section 7.3 has a further

Table 7.7. Type III Analysis of Variance for Responses of Example 7.2.

Source of variation	Degrees of freedom	Sum of squares	Mean square
Total	124	5076.000	—
Correction for mean	1	4512.129	—
Replicate R	3	4.229	1.410
Tillage T	3	22.668	7.556
$R \times T =$ error A	9	10.854	1.206
Genotype G	21	309.908	14.758
$G \times T$	63	95.140	1.510
$C \times R$ within T	24	25.167	1.049

Table 7.8. Least Squares Means for Tillage and Genotype.

Genotype	T1	T2	T3	T4	Genotype
1	8.92	8.92	3.83	3.08	6.19
2	7.92	7.92	4.83	4.08	6.19
3	6.92	4.92	5.83	5.08	5.69
4	5.92	4.92	2.83	2.08	3.94
5	9.92	7.92	4.83	4.08	6.69
6	2.58	2.92	3.83	4.75	3.52
7	2.58	1.92	3.83	3.75	3.02
8	3.58	1.91	3.83	4.75	3.52
9	3.38	4.92	6.83	6.75	5.52
10	8.58	6.92	7.83	6.75	7.52
11	8.92	9.58	7.50	6.75	8.19
12	8.92	7.58	7.50	6.75	7.69
13	0.92	2.58	3.50	4.75	2.94
14	1.92	1.58	2.50	2.75	2.19
15	8.92	9.58	8.50	7.75	8.69
16	8.58	7.58	6.83	5.42	7.10
17	8.58	8.58	6.83	5.42	7.35
18	8.58	6.58	8.83	6.41	7.60
19	8.58	5.58	8.83	7.42	7.60
20	7.00	8.50	6.50	5.50	6.88
21	5.75	5.00	5.00	4.25	5.00
22	7.00	8.25	7.00	6.50	7.19
Tillage	6.55	6.10	5.80	5.22	6.03

partitioning of the split plot experimental units into split split plot experimental units to which the split split plot treatments are applied. There are n/r new genotypes in each replicate. A partitioning of the degrees of freedom in an analysis of variance table for such an experiment design is

Source of variation	Degrees of freedom
Total	$rtcf + tfn$
Correction for mean	1
Replicate R	$r - 1$
Whole plot treatments W	$t - 1$
$R \times W =$ error W	$(r - 1)(t - 1)$
Split plot treatments S	$v - 1$
$W \times S$	$(t - 1)(v - 1)$
$\quad W \times$ checks C	$(t - 1)(c - 1)$
$\quad W \times$ new, N	$(t - 1)(n - 1)$
$\quad W \times C$ versus N	$t - 1$
$S \times R$ within W	$t(c - 1)(r - 1)$
Split split plot treatments SS	$f - 1$

(*Continued*)

(*Continued*)

Source of variation	Degrees of freedom
$SS \times W$	$(t-1)(f-1)$
$SS \times S$	$(f-1)(v-1)$
$SS \times W \times S$	$(f-1)(v-1)(t-1)$
$SS \times R$ within W & S	$tc(f-1)(r-1)$

If desired, the interactions of S and SS for two factors may be partitioned in a similar manner as for the $W \times S$ interaction.

Table 7.9. Responses for the ASSPED.

Replicate 1

	T1								T2						
	Genotype								Genotype						
Fertilizer	1	2	3	4	5	21	22	23	Fertilizer	1	2	3	4	5	21	22	23
F1	2	3	7	8	8	2	4	6	F1	3	4	7	9	7	3	5	6
F2	4	5	7	8	8	5	6	7	F2	4	6	8	8	9	6	7	7
F3	6	7	9	9	9	7	7	8	F3	5	8	9	9	9	7	8	8

Replicate 2

	T1								T2							
	Genotype								Genotype							
Fertilizer	6	7	8	9	10	21	22	23	Fertilizer	6	7	8	9	10	21	22	23
F1	4	6	7	7	3	3	5	6	F1	5	6	7	7	4	2	6	6
F2	5	7	8	8	5	5	5	7	F2	6	7	9	7	5	4	7	7
F3	6	8	9	9	6	7	7	9	F3	7	7	8	9	7	5	8	8

Replicate 3

	T1								T2							
	Genotype								Genotype							
Fertilizer	11	12	13	14	15	21	22	23	Fertilizer	11	12	13	14	15	21	22	23
F1	4	7	6	7	2	4	6	7	F1	5	8	7	7	4	3	7	7
F2	5	8	7	7	3	5	7	8	F2	5	8	7	8	5	6	7	8
F3	6	8	7	9	4	5	8	8	F3	6	9	9	8	7	6	9	9

Replicate 4

	T1								T2							
	Genotype								Genotype							
Fertilizer	16	17	18	19	20	21	22	23	Fertilizer	16	17	18	19	20	21	22	23
F1	1	2	3	4	5	5	5	6	F1	2	3	4	5	6	4	7	7
F2	3	4	4	5	5	6	7	8	F2	2	5	6	7	7	6	8	8
F3	5	5	6	7	7	8	7	9	F3	4	6	7	7	8	7	8	9

Table 7.10. Type III Analysis of Variance for Responses in Table 7.9.

Source of variation	Degrees of freedom	Sum of squares	Mean square
Total	192	8205.000	—
Correction for mean	1	7537.547	—
Replicate R	3	12.486	4.162
Tillage T	1	11.158	11.158
$T \times R$	3	1.153	0.384
Genotype G	22	389.007	17.682
$G \times T$	22	18.424	0.837
$G \times R$ within T	12	8.778	0.731
Fertilizer F	2	120.563	60.282
$F \times T$	2	0.822	0.411
$F \times G$	44	23.459	0.533
$F \times G \times T$	44	10.740	0.244
$F \times R$ within G & T	36	11.333	0.315

Example 7.3. To illustrate the analysis for the above experiment design, let the number of replicates or complete blocks be $r = 4$, let the number of whole plot treatments, tillage, be $t = 2$, let the number of split plot treatments, genotypes, in each whole plot be $(c = 3$ checks$) + (n/r = 20/4 = 5$ new genotypes$) = 3 + 5 = 8$, and let the number of split split plot treatments, fertilizer, be $f = 3$. The fertilizer treatments may be applied as described for Example 7.1. The new treatments are numbered from 1 to 20 and the checks are numbered 21, 22, and 23. Hypothetical data for an experiment designed as an ASSPED are given in Table 7.9.

An analysis of variance table for the data in Table 7.9 is given in Table 7.10. Further partitioning of the degrees of freedom and sums of squares as described above and for Example 7.1 may be made here as well. F-test statistics may be added to the following table if desired.

Table 7.11. Estimable Least Squares Means for Data of Table 7.9.

Three factor means							
		T1				T2	
Check	F1	F2	F3	Check	F1	F2	F3
21	3.50	5.25	7.75	21	3.00	5.50	6.25
22	5.00	6.25	7.25	22	6.25	7.25	8.25
23	6.25	7.25	8.50	23	6.50	7.50	8.50
Check by fertilizer means				Check by tillage means			
Check	F1	F2	F3	Check	T1	T2	Check mean
21	3.25	5.37	6.50	21	5.17	4.92	5.04
22	5.62	6.75	7.75	22	6.17	7.25	6.71
23	6.37	7.50	8.50	23	7.42	7.50	7.46

Only least squares check means are given by SAS. Since least squares means are not estimable, solutions for the effects are obtained and are given by the computer code in Appendix 7.3. There are a large number of solutions involving all of the effects. They may be found in Federer and Arguillas (2005a). The least squares means that are estimable are given in Table 7.11.

7.5. DISCUSSION

As previously indicated, there are many variations of the ASPED. A combination of split plot and split block experiment designs may be required by the experimental conditions. For example, an augmented split block design for two factors like tillage T and fertilizer F could be combined with genotypes G as split plots to either T or F or T and F. Factors F and G could be in a split block arrangement (Federer, 2005) with T as a split plot treatment. Another variation would be to have a split split split plot experiment design for T, G, F, and a fourth factor, say herbicides H. Instead of using augmented randomized complete block designs, use may be made of other augmented designs such as an augmented Latin square or rectangle design, an augmented incomplete block design, an augmented lattice square design (Federer, 2002), and others. These designs could be used in combination with split block and split plot arrangements, should the experimenter prefer such arrangements.

The randomization procedure for these experiment designs follows that for the nonaugmented experiment designs. An exception would be for the parsimonious designs presented by Federer and Scully (1993). This is discussed in Section 7.2. It would not be advisable to have adjacent levels largely different for some factors, say fertilizer treatments, adjacent to each other because of inter-plot competition. That is, quantitative levels would be placed adjacent to each other but in decreasing values of the level of a factor. Spacing or bordering the different levels reasonably far apart, about the height of the plant, may avoid inter-plot competition. The density within a plot could be increased to have constant density per hectare for different plant or row spacing treatments.

More detailed SAS computer outputs for the three numerical examples presented here may be found in Federer and Arguillas (2005a).

7.6. PROBLEMS

Problem 7.1. Suppose that the experimenter eliminated genotypes 4, 7, and 9 from consideration in Example 7.1. Obtain the analysis as described in Section 7.2 for the remainder of the experiment. Compare your results with those presented for Example 7.1.

Problem 7.2. Suppose that the experimenter decided not to obtain responses for genotypes 15, 16, and 17 in Example 7.3 because of their poor performance and appearance in the field. Perform an analysis for the remainder of the data. Compare your results with those given for Example 7.3.

7.7 REFERENCES

Federer, W. T. (1956). Augmented (or hoonuiaku) designs. *Hawaiian Planters' Record* 55:195–208.

Federer, W. T. (1961). Augmented designs with one-way elimination of heterogeneity. *Biometrics* 17:447–473.

Federer, W. T. (1993). *Statistical Design and Analysis for Intercropping Experiments, Volume I: Two Crops.* Springer-Verlag, New York, Berlin, Heidelburg, London, Paris, Tokyo, Hong Kong, Barcelona, Budapest, Chapter 10.

Federer, W. T. (1998). Recovery of interblock, intergradient, and intervariety information in incomplete block and lattice rectangle designed experiments. *Biometrics* 54:471–481.

Federer, W. T. (2002). Construction and analysis for an augmented lattice square experiment design. *Biometrical J* 44:261–257.

Federer, W. T. (2003). Analysis for an experiment designed as an augmented lattice square design. In *Handbook of Formulas and Software for Plant Geneticists and Breeders* (M. S. Kang, Editor), Food Products Press, Binghamton, New York, Chapter 27, pp. 283–289.

Federer, W. T. (2005). Augmented split block experiment design. *Agronomy J* 97:578–586.

Federer, W. T. and F. O. Arguillas, Jr. (2005a). Augmented split plot experiment design. BU-1658-M in the Technical Report Series of the Department of Biological Statistics and Computational Biology (BSCB), *Cornell University*, Ithaca, NY 14853, June.

Federer, W. T. and F. O. Arguillas, Jr. (2005b). Augmented split plot experiment design. *J Crop Improvement* 15(1):81–96.

Federer, W. T., R. C. Nair, and D. Raghavarao. (1975). Some augmented row-column designs. *Biometrics* 31:361–373.

Federer, W. T. and D. Raghavarao. (1975). On augmented designs. *Biometrics* 31:29–35.

Federer, W. T. and B. T. Scully. (1993). A parsimonious statistical design and breeding procedure for evaluating and selecting desirable characteristics over environments. *Theor App Genet* 86:612–620.

Federer, W. T. and R. D. Wolfinger. (2003). Augmented row-column designs and trend analyses. In *Handbook of Formulas and Software for Plant Geneticists and Breeders* (M. S. Kang, Editor), Food Products Press, Binghamton, New York, Chapter 28, pp. 291–295.

SAS Institute, Inc. (1999–2001). Release 8.02, copyright, Cary, NC.

Wolfinger, R. D., W. T. Federer, and O. Cordero-Brana. (1997). Recovering information in augmented designs, using SAS PROC GLM and PROC MIXED. *Agronomy J* 89:856–859.

APPENDIX 7.1. SAS CODE FOR ASPED, GENOTYPES AS WHOLE PLOTS, EXAMPLE 7.1

A computer code for an analysis of the data in Table 7.1 is presented below. The data set is given in an unabbreviated form in order to demonstrate the data entry into the data file. These data also appear on the book's FTP site.

```
data asped;
input Y R G F;
if (G > 24) then new = 0; else new = 1;
if (new) then Gn = 999; else Gn = G;
datalines;
2   1   25   1
4   1   25   2
6   1   25   3
1   1   26   1
3   1   26   2
5   1   26   3
9   1   27   1
9   1   27   2
8   1   27   3
9   1   28   1
9   1   28   2
7   1   28   3
2   1    1   1
5   1    1   2
7   1    1   3
3   1    2   1
6   1    2   2
5   1    2   3
4   1    3   1
7   1    3   2
6   1    3   3
5   1    4   1
8   1    4   2
4   1    4   3
6   1    5   1
8   1    5   2
8   1    5   3
7   1    6   1
8   1    6   2
7   1    6   3
3   2   25   1
3   2   25   2
7   2   25   3
2   2   26   1
2   2   26   2
4   2   26   3
```

APPENDIX

8	2	27	1
8	2	27	2
8	2	27	3
7	2	28	1
8	2	28	2
9	2	28	3
1	2	7	1
2	2	7	2
3	2	7	3
2	2	8	1
3	2	8	2
5	2	8	3
3	2	9	1
4	2	9	2
4	2	9	3
4	2	10	1
4	2	10	2
5	2	10	3
8	2	11	1
8	2	11	2
8	2	11	3
3	2	12	1
5	2	12	2
7	2	12	3
4	3	25	1
6	3	25	2
8	3	25	3
2	3	26	1
5	3	26	2
7	3	26	3
8	3	27	1
7	3	27	2
9	3	27	3
7	3	28	1
7	3	28	2
9	3	28	3
7	3	13	1
7	3	13	2
9	3	13	3
5	3	14	1
6	3	14	2
8	3	14	3
3	3	15	1
5	3	15	2
6	3	15	3
7	3	16	1
7	3	16	2
9	3	16	3
6	3	17	1

```
8  3  17  2
8  3  17  3
5  3  18  1
7  3  18  2
8  3  18  3
4  4  25  1
5  4  25  2
6  4  25  3
5  4  26  1
2  4  26  2
5  4  26  3
9  4  27  1
9  4  27  2
9  4  27  3
9  4  28  1
8  4  28  2
7  4  28  3
5  4  19  1
8  4  19  2
9  4  19  3
6  4  20  1
6  4  20  2
8  4  20  3
7  4  21  1
4  4  21  2
8  4  21  3
8  4  22  1
7  4  22  2
9  4  22  3
9  4  23  1
8  4  23  2
9  4  23  3
9  4  24  1
8  4  24  2
9  4  24  3
;

ods trace on;
proc glm data = asped;
class R G F;
model Y = R G R*G F F*G/solution ;
lsmeans F G F*G/ out = solution noprint;
ods output parameterestimates = aspedparameterestimates;
run;
/*Only G check means and check by F means are estimable*/
ods trace off;
proc sort data = aspedparameterestimates;
   By descending Estimate ;
proc print; run;
```

APPENDIX   **185**

```
odslistingclose;   /*Allows unordered solutions to not be printed.*/
ods trace on;
proc mixed data = asped;
  class R F G Gn;
  model Y = F Gn F*Gn/solution;
  random R R*Gn G*new/solution;
  lsmeans Gn F F*Gn;
ods output solutionr = solutionforrandomeffects;
ods output lsmeans = leastsquaresmeans;
run;
ods trace off;
ods listing;   /* End of no printing statement of solutions.*/
proc sort data = solutionforrandomeffects;
  By descending Estimate;
Proc print; run;
proc sort data = leastsquaresmeans;
  By descending Estimate;
proc print; run;
```

APPENDIX 7.2. SAS CODE FOR ASPEDT, GENOTYPES AS SPLIT PLOTS, EXAMPLE 7.2

The code for Example 7.2 is presented below. The code allows for the least squares means, lsmeans, to be ranked in descending order of the response variable Y. For this experiment design, all means are estimable for fixed effects. The complete data set is given on the book's FTP site.

```
data aspedt;

input Y R T G;
If (G > 19) then new = 0; else new = 1;
If (G < 20) then Gn = 999; else Gn = G;
datalines;

6    1    1    20
4    1    1    21
.... ...
7    4    4    18
8    4    4    19
;

proc glm data = aspedt;
class R T G;
model Y = R T R*T G G*T;
lsmeans T G T*G/out = lsmeans noprint;
/*All means estimable in Example 3.2.*/
run;
```

```
proc sort data = lsmeans;
  By descending LSMEAN;
proc print; Run;
ods trace on;
proc mixed data = aspedt;
  class R G T Gn;
  model Y = T Gn Gn*T/solution;
  random R R*T G*new/solution;
  lsmeans Gn T Gn*T;
ods output solutionf = solutionforfixedeffects;
ods output solutionr = solutionforrandomeffects;
ods output lsmeans = leastsquaresmeans;
run;

ods trace off;
proc sort data = solutionforfixedeffects;;
  By descending Estimate;
proc print; run;
proc sort data = solutionforrandomeffects;
  By descending Estimate;
proc print; run;
proc sort data = leastsquaresmeans;
  By descending Estimate;
proc print; run;
```

APPENDIX 7.3. SAS CODE FOR ASSPED, EXAMPLE 7.3

The code for Example 7.3 is given below. The least squares means for other than checks and interactions of T and F with checks, are nonestimable. The complete data set is given on the book's FTP site.

```
data assped;
input Y R T G F;
if (G > 20) then new = 0; else new = 1;
if (new) then Gn = 999; else Gn = G;
datalines;

2   1   1    1    1
4   1   1    1    2
...
8   4   2   23    2
9   4   2   23    3
;

ods trace on;
proc glm data = assped;
class R T G F;
```

APPENDIX

```
model Y = R T R*T G G*T R*G(T) F F*T F*G F*G*T/solution;
  /*only means estimable are check G means*/
lsmeans T G T*G F F*T F*G F*G*T/out = solution;
ods output parameterestimates = asspedparameterestimates;
ods trace off;
proc sort data = asspedparameterestimates;
  by descending Estimate;
proc print; run;
ods listing close;   /*Unordered solutions not printed.*/
ods trace on;
proc mixed data = assped;
  class R T G F Gn;
  model Y = T Gn Gn*T F F*T F*Gn F*Gn*T/solution;
  random R R*T Gn*R(T) G*new/solution;
  lsmeans T Gn Gn*T F F*Gn F*T F*Gn*T;
ods output solutionf = solutionforfixedeffects;
ods output solutionr = solutionforrandomeffects;
ods output lsmeans = leastsquaresmeans;
run;

ods trace off;
ods listing;   /*End of no printing statement.*/
proc sort data = solutionforfixedeffects;
  by descending Estimate;
proc print; run;
proc sort data = solutionforrandomeffects;
  by descending Estimate;
proc print; run;
proc sort data = leastsquaresmeans;
  by descending Estimate;
proc print; run;
```

CHAPTER 8

Augmented Split Block Experiment Design

8.1. INTRODUCTION

A new class of augmented experiment designs as described by Federer (2005) is presented herein. This class of experiment designs will allow screening treatments from each of two or more factors simultaneously, or screening the elements of one factor, for example genotypes, under a variety of conditions. This class demonstrates another aspect of the flexibility available to an experimenter as compared to the classes of augmented experiment designs available before this paper (see references in Chapter 7). This class of experiment designs bears some similarities to the parsimonious designs discussed by Federer and Scully (1993) and Federer (1993), and the class of split block designs with controls presented by Mejza (1998). Some members of the class of augmented split block experiment designs are presented in the next section using examples to illustrate the construction of these designs. In Section 8.3, the use of these designs in the context of intercropping experiments is discussed. A numerical example is presented in Section 8.4 to demonstrate the types of analyses that accompany these designs. A few comments on these designs are presented in the last section.

8.2. AUGMENTED SPLIT BLOCK EXPERIMENT DESIGNS

The ideas in the above references are used to construct *augmented split block experiment designs*. A linear model for each of the augmented experiment designs is

---

Variations on Split Plot and Split Block Experiment Designs, by Walter T. Federer and Freedom King
Copyright © 2007 John Wiley & Sons, Inc.

given by the sources of variation listed in the partitioning of the total degrees of freedom in the ANOVA table. Illustrations are used to demonstrate the construction and analysis of this class of designs. For example, suppose that an experimenter used $ac = 4$ standard or control genotypes and $an = 64$ new genotypes as the A factor and two levels of fertilizers in combination with two types of soil preparation as the B factor. Likewise, it may be that this experiment is to be conducted at several sites or locations that have certain characteristics. It is desired to test or screen the new treatments on the $b = 4$ factor B treatments, B1, B2, B3, and B4. Further suppose that the experiment designs for the factor A and for the factor B were an augmented randomized complete block design and a randomized complete block experiment design, respectively. The four factor B treatments are in a split block arrangement to the eight factor A treatments in each block. Given that eight complete blocks, replicates, are to be used and that eight of the 64 new factor A treatments are to be included in each of the eight blocks as the augmented treatments, an analysis of variance table with a partitioning of the degrees of freedom is given below:

	Degrees of freedom	
Source of variation	Example	General
Total	384	$rb(ac + an/r)$
Correction for the mean	1	1
Replicate $= R$	7	$r - 1$
A treatments	67	$ac + an - 1 = v - 1$
Control, C	3	$ac - 1$
Control versus new	1	1
New	63	$an - 1$
$C \times R =$ error A	21	$(ac - 1)(r - 1)$
B	3	$b - 1$
Soil	1	$s - 1$
Fertilizer	1	$f - 1$
Soil × fertilizer	1	$(s - 1)(f - 1)$
$B \times R$	21	$(b - 1)(r - 1)$
$A \times B$	201	$(a - 1)(b - 1)$
Control × B	9	$(ac - 1)(b - 1)$
Control vs. new × B	3	$(1)(b - 1)$
New × B	189	$(an - 1)(b - 1)$
$C \times B \times R$	63	$(ac - 1)(b - 1)(r - 1)$

As a second example, suppose that the conditions of the above arrangement hold except that an incomplete block design with r incomplete blocks of size $k = 2$ is used for the $ac = 4$ factor A check treatments, A1, A2, A3, and A4. A plan before randomization and before including the augmented treatments for this example is of the form:

Block 1	Block 2	Block 3	Block 4
A1 A2	A2 A3	A3 A4	A4 A1
B1	B1	B1	B1
B2	B2	B2	B2
B3	B3	B3	B3
B4	B4	B4	B4

In the above incomplete block arrangement for factor A check treatments, increase the block size to eighteen experimental units. Sixteen of the 64 new factor A treatments would be included in each of the four blocks. A set of sixteen new and two controls or checks would be randomly allotted to the eighteen experimental units for factor A in each of the four incomplete blocks. The B treatments would run across these eighteen factor A treatments in a split block manner in each of the $r = 4$ blocks. The B treatments are in a randomized complete block design arrangement.

A linear model may be obtained from the partitioning of the degrees of freedom for the above designed experiment as follows:

Source of variation	Degrees of freedom	
	Example	General
Total	288	$anb + rbac$
Correction for mean	1	1
Block	3	$r - 1$
A treatments	67	$ac + an - 1 = a - 1$
Control or check, C	3	$ac - 1$
Control versus new	1	1
New	63	$an - 1$
$C \times$ block = intrablock error for A	1	$rk - ac - r + 1$
B	3	$b - 1$
$B \times$ block = error for B	9	$(b - 1)(r - 1)$
$A \times B$	201	$(a - 1)(b - 1)$
$B \times$ control	9	$(ac - 1)(b - 1)$
$B \times$ control versus new	3	$b - 1$
$B \times$ new	189	$(b - 1)(an - 1)$
$C \times B \times$ block = error for $A \times B$	3	$(rk - ac - r + 1)(b - 1)$

Note that the $C \times$ block, the error term for factor A, has only one degree of freedom as the block sum of squares eliminating checks is partitioned into the incomplete blocks within replicate with two degrees of freedom, and the remaining one for the Error A which is the intrablock error from an incomplete block experiment design.

Though the above design is connected, there are insufficient degrees of freedom associated with the error terms. Hence, additional blocks and/or additional checks in a block will be required. The three interactions with block mean squares would form the error terms for this augmented split block experiment design. If more new treatments are to be tested, the number of blocks could be increased and more new treatments could be included in the blocks of the factor A experimental units.

As a further variation of an augmented split block experiment design, consider the following arrangement where both factors are designed as augmented randomized complete blocks:

Block 1 Factor A Factor B 1 2 3 4 5 6 7	Block 2 Factor A Factor B 1 2 3 4 8 9 10	Block 3 Factor A Factor B 1 2 3 4 11 12 13
1	1	1
2	2	2
3	3	3
4	6	8
5	7	9

The control treatments for factor A are 1, 2, 3, and 4 and the new or augmented factor A treatments are numbered 5 to 13. The number of check or control factor A treatments is defined as $ac = 4$ and the number of new factor A treatments as $an = 9$. The augmented or new treatments are included only once in the experiment. For the B factor, numbers 1, 2, and 3, are the control treatments and $bc = 3$ is defined to be the number of checks. Numbers 4 to 9 are the new factor B treatments and we let $bn = 6$ be the number of new factor B treatments. An analysis of variance partitioning of the 105 degrees of freedom follows:

	Degrees of freedom	
Source of variation	Example	General
Total	$3(5)(7) = 105$	$r(ac + an/r)(bc + bn/r)$
Correction for the mean	1	1
Block = R	2	$r - 1$
A	12	$ac + an - 1$
Control = AC	3	$ac - 1$
New = AN	8	$an - 1$
AC versus AN	1	1
$AC \times R$ = error A	$2(3) = 6$	$(ac - 1)(r - 1)$
B	8	$bc + bn - 1$
Control = BC	2	$bc - 1$
New = BN	5	$bn - 1$
BC versus BN = BCN	1	1
$BC \times R$ = error B	$2(2) = 4$	$(bc - 1)(r - 1)$
$A \times B$	60	$ac(bc + bn - 1)$ $+ (an - 1)(bc - 1) + bn(an/r - 1)$
$AC \times BC$	$3(2) = 6$	$(ac - 1)(bc - 1)$
$AC \times BN$	$3(5) = 15$	$(ac - 1)(bn - 1)$

(*Continued*)

(*Continued*)

Source of variation	Degrees of freedom	
	Example	General
$AC \times BC$ vs. AN	$3(1) = 3$	$(ac-1)(1)$
$AN \times BC$	$8(2) = 16$	$(an-1)(bc-1)$
$AN \times BN$/block	$2(1)(3) = 6$	$r(an/r - 1)(bn/r - 1)$
$AN \times BC$ vs. BN within block	$2(1)(3) = 6$	$r(1)(an/r - 1)$
AC vs. $AN \times BC$	$1(2) = 2$	$1(bc-1)$
AC vs. $AN \times BN$ within block	$1(1)(3) = 3$	$1(bn/r - 1)(r)$
AC vs. $AN \times BC$ vs. BN within block	$1(1)(3) = 3$	r
$AC \times BC \times R =$ error AB	$3(2)(2) = 12$	$(ac-1)(bc-1)(r-1)$

Note that not all interactions of new treatments are obtained. The interactions of new treatments in the same block are obtainable as there are two new B treatments and three new A treatments in any one block. For some of the interactions, the contrasts vary from block to block and will be different. This is why some of the degrees of freedom were partitioned by block.

Suppose the blocks for factor A were located in adjacent positions, that is, side by side. If the nine B treatments were to go across all three blocks of the previous example and the A treatments stayed the same, then there would be a total $9(4 + 3)(3) = 189$ experimental units. The linear model is obtained from the partitioning of the degrees of freedom given below:

Source of variation	Degrees of freedom
Total	189
Correction for the mean	1
Block $= R$	2
Factor $A =$ error A	12
Control $= AC$	3
New $= AN$	8
AC versus AN	1
$AC \times R =$ Error A	$2(3) = 6$
Factor B	8
BC	2
BN	5
BC vs. BN	1
$B \times R$ (not an error term)	$8(2) = 16$
$B \times A$	$8(12) = 96$
$B \times AC$	$8(3) = 24$
$B \times AN$	$8(8) = 64$
$B \times AC$ vs. AN	$8(1) = 8$
$A \times B \times R = AC \times B \times R =$ error AB	$3(8)(2) = 48$

For the above arrangement, all $A \times B$ interaction terms are available. There is no error term for factor B as there is only one layout (randomization) for this factor.

8.3. AUGMENTED SPLIT BLOCKS FOR INTERCROPPING EXPERIMENTS

Consider the penultimate arrangement in the previous section. Here the treatments 1, 2, 3, and 4 of factor A could be four standard maize cultivars and factor A treatments 5 to 13 could be nine promising new genotypes of maize. Factor B treatments 1, 2, and 3 could be three standard bean cultivars and treatments 4 to 9 could be six promising new bean genotypes. The goal could be to determine how these mixtures of maize and beans perform as an intercrop. The maize could be planted in rows and the rows of beans could be planted perpendicular to the maize rows, in each block. However, not all possible mixtures result here. If this is desired, then the last example of the previous section could be used.

In intercropping and other mixture experiments, responses may be available for the mixture or for individual components of the mixture. If the latter responses are available, analyses may be conducted for each of the components. For example in intercropping experiments with maize and beans, responses for maize and for beans may be obtained. These responses may also be combined and an analysis conducted on the combined responses. When responses are available for each crop in the mixture, it is possible to estimate the various mixing effects for each component of the mixture (See Federer, 1993, 1999).

Suppose that an experimenter desired to screen one set of new genotypes, say beans, and another set of new genotypes, say maize, to determine their suitability for intercropping systems. The treatments for the A factor could be the controls and new genotypes of maize and the treatments for factor B could be the controls and new genotypes for beans. Further, suppose that an augmented randomized complete block experiment design was used for both the A and B factors with $r = 8$ blocks. Let the number of new maize genotypes be 100 and the number of new bean genotypes be 96. Twelve new bean genotypes with the four bean controls are randomly allotted to the 16 factor B experimental units in each of the $r = 8$ blocks, replicates. Given that five maize genotypes are the controls, there will be $5 + 12 = 17$ factor A experimental units in four of the blocks and $5 + 13 = 18$ experimental units in the other four blocks to accommodate the 100 new maize genotypes. These would be randomly allotted to each of the eight blocks. There is a total of $4(16)(17) + 4(16)(18) = 2{,}240$ experimental units. A partitioning of the degrees of freedom in an analysis of variance for this situation is:

Source of variation	Degrees of freedom
Total	$[4(5 + 12) + 4(5 + 13)][4 + 12] = 2{,}240$
Correction for mean	1
Replicate $= R$	7

(*Continued*)

(*Continued*)

Source of variation	Degrees of freedom
Maize genotypes = A	104
Maize control = MC	4
MC versus new = MCN	1
New = MN	99
$MC \times R$	$4(7) = 28$
Bean genotypes	99
Bean control = BC	3
BC versus new = BCN	1
New = BN	95
$BC \times R$	$3(7) = 21$
$A \times B$	1896
$BC \times MC$	$3(4) = 12$
$BCN \times MC$	$1(4) = 4$
$BN \times MC$	$95(4) = 380$
$BC \times MCN$	$3(1) = 3$
$BN \times MCN$ within block	$11(1)(8) = 88$
$BCN \times MCN$ within block	$1(1)(8) = 8$
$BC \times MN$	$3(99) = 297$
$BCN \times MN$ within block	$4(1)(11) + 4(1)(12) = 92$
$BN \times MN$ within block	$4(11)(11) + 4(11)(12) = 1{,}012$
$AC \times BC \times R$	84

8.4. NUMERICAL EXAMPLE 8.1

To illustrate the type of effects that are estimable from an augmented split block experiment design, a hypothetical numerical example was constructed. The number of factor A treatments is $ac + an = 13 = 4$ controls $+9$ new treatments. The checks are numbered A1, A2, A3, and A4 and the new treatments are numbered A5 to A13. The number of factor B treatments is $bc + bn = 9 = 3$ controls $+6$ new treatments. The controls are numbered B1, B2, and B3 and the new are numbered B4 to B9. An augmented randomized complete block design with $r = 3$ blocks is used for each of the factor A and factor B treatments. A systematic layout of the design and the responses (data) is given in Table 8.1.

An analysis of variance along with F-values is presented in Table 8.2. The sums of squares for the new treatments and the contrast of the new treatments versus the control treatments were pooled in the table. It is possible to obtain the sums of squares for each of the effects as given above, but that was not done for the example.

To obtain the analysis of variance table, three runs of SAS PROC GLM and one run of SAS PROC MIXED were conducted. These four runs were for the A controls and the B treatments, the B controls and the A treatments, for all 105 observations, and for random new genotype effects. It is possible to obtain a further partitioning of the sums of squares and degrees of freedom. This partitioning was described in the previous section.

NUMERICAL EXAMPLE 8.1

Table 8.1. Systematic Layout and Artificial Responses for an ASBED.

	Block 1						Block2						Block3				
	B1	B2	B3	B4	B5		B1	B2	B3	B6	B7		B1	B2	B3	B8	B9
A1	9	7	5	9	1	A1	9	7	5	9	11	A1	11	10	10	15	9
A2	3	5	4	9	7	A2	13	5	4	9	7	A2	13	15	14	9	17
A3	8	7	5	9	2	A3	8	7	5	9	12	A3	8	17	15	9	12
A4	3	5	4	9	7	A4	13	5	4	9	7	A4	13	15	14	9	17
A5	8	7	5	9	2	A8	13	8	8	9	7	A11	13	15	14	19	17
A6	3	5	4	9	7	A9	8	7	8	9	12	A12	18	17	15	19	12
A7	8	7	5	9	2	A10	13	9	12	9	7	A13	13	15	14	19	17

Using the following SAS PROC GLM MODEL statement

Response = R A R*A B B*R A*B/solution;

it is possible to obtain solutions for all effects that are present in the model. These solutions for the effects are useful as the LSMEANS statement does not produce means for the new treatments. The effects obtained are presented in Table 8.3. No solutions are available for the empty spaces in the table. The sum of squares for new versus checks is pooled with the new sum of squares for six degrees of freedom. This is because there are combinations of new treatments that do not appear in a block. Solutions are possible only for combinations that appear in the design. When interpreting these values, the procedure used to obtain them must be well

Table 8.2. Type III Analysis of Variance for the Data in Table 8.1.

Source of variation	Degrees of freedom	Sum of squares	Mean square	F-value
Block = R	2	372.22	186.11	
Factor A	12	93.00	7.75	0.93
Control = AC	3	2.75	0.92	0.11
New = AN +	9	90.25	10.03	1.20
AC vs. AN = ACN				
$A \times R$	6	50.00	8.33	
B	8	123.79	15.47	0.71
Control = BC	2	21.56	10.78	0.49
New = BN +	6	102.23	17.04	0.78
BC vs. BN = BCN				
$B \times R$	4	87.44	21.86	
$A \times B$	60	380.83	6.35	1.56
$AC \times BC$	6	18.00	3.00	0.74
$AC \times BN + AC \times BCN$	18	150.80	8.38	2.05
$AN \times BC + ACN \times BC$	18	42.33	2.35	0.58
Other interactions	18	169.70	9.43	2.31
$AC \times BC \times R$	12	49.00	4.08	

Table 8.3. SAS Solutions for Effects of the Responses of Table.8.1.

A treatments	\multicolumn{9}{c}{B treatments}	A effect
	1	2	3	4	5	6	7	8	9	
1	4.44	0.50	0.00	0.00	0.00	0.00	0.00	0.00	0.00	−8.00
2	−0.22	−0.05	0.28	2.25	7.25	7.58	7.58	−10.00	0.00	−0.00
3	2.00	5.83	5.17	3.92	3.92	12.58	17.58	−5.00	0.00	−5.00
4	−0.22	−0.06	0.28	2.25	7.25	7.58	7.58	−10.00	0.00	−0.00
5	0.75	0.50	−0.00	−0.00	0.00					2.25
6	−9.25	−6.50	−6.00	−5.00	0.00					7.25
7	0.75	0.50	0.00	0.00	0.00					2.25
8	−6.25	−5.50	−4.00			0.00	0.00			7.58
9	−16.25	−11.50	−9.00			−5.00	0.00			12.58
10	−6.25	−4.50	0.00			0.00	0.00			7.58
11	0.00	0.00	−0.00					0.00	0.00	−0.00
12	10.00	7.00	6.00					0.00	0.00	−5.00
13	0.00	0.00	0.00					0.00	0.00	0.00
B tr. Effect	−4.00	−2.00	−3.00	0.00	0.00	0.00	0.00	2.00	0.00	

understood. The SAS PROC GLM procedure does not use the constraint that the sum of the effects for a factor is zero. Instead, the procedure uses the constraint that the highest numbered effect is set equal to zero. The consequence of this is that the highest numbered effect for factor A (and for factor B) is subtracted from each of the other factor A effects. For interactions, the effects in the last row and in the last column are set to zero. That is, the last row effect is subtracted from each of the other effects in a row and likewise for columns.

The SAS PROC GLM codes for the above analyses are given in Appendix 8.1.

The least squares means that appear in the SAS output are given in Table 8.4.

The means for the new treatments are obtained as the response for a new treatment minus the block effect in which the new treatment occurred. The block

Table 8.4. Least Squares Means for the A and B Control Treatments.

A controls	B controls			Mean
	1	2	3	
1	9.67	8.00	6.67	8.11
2	9.67	8.33	7.33	8.44
3	8.00	10.33	8.33	8.89
4	9.67	8.33	7.33	8.44
Mean	9.25	8.75	7.42	8.47

effects may be obtained from an analysis of the control data only. The new treatments do not contribute to estimating the block effects as they appear only once in the experiment (Federer and Raghavarao, 1975).

8.5. COMMENTS

Any experiment design may be augmented to accommodate a set of new treatments that are to be replicated one or more times, but usually once. The class of augmented experiment designs is large and varied. Lack of material may be a factor in deciding to use only one replicate. Or, it may be that the experimenter has so many new lines that makes it undesirable or unrealistic to use more than one replicate of each of the new treatments. For example, plant breeders of some crops may want to screen up to 30,000 new entries each year. Others screen up to 8000 new entries each year. Producers of fungicides, herbicides, and others may have hundreds of new entries for screening. Since some of these treatments may kill all the plants in an experimental unit, it is not desirable to undertake screening on more than one experimental unit.

In field experiments involving intercrops, it is possible to first plant one crop in the field and then, perpendicular to the planting of crop one, a second crop is planted across the first crop in each of the blocks, or possibly across the entire experiment. Such plans as the arrangements discussed above, allow the experimenter to put the new treatments of two crops in each of the blocks in the desired arrangement to obtain two crop mixing or combining effects for the selected combinations. For three crops in a mixture, an experimenter could use one crop as the whole plot of a split plot design and then use an augmented split block design, as described above, for the other two factors or crops as the split plot treatments.

As stated, the use of an augmented split block experiment design allows the experimenter to screen new treatments for various cultural or management practices. The new treatments used in this situation would more often be new treatments that have survived previous stages of screening and may be considered to be fixed rather than random effects. Cultural practices such as soil preparation would be ideal for this design. Spraying fungicides or herbicide treatments would also fit into this design as would fertilizer levels, density, time of planting, spacing of plants, time of spraying, and so forth. These designs are useful for any type of experiment that involves the screening of material, where one or more conditions for screening are desired.

8.6. PROBLEMS

Problem 8.1. For the data in Table 8.1, omit the results for B6 and B7. Obtain an analysis of the data for this set. Compare your results with those obtained in the text for the complete data set.

Problem 8.2. For the data in Table 8.1, omit the results for A7 and A10. Obtain an analysis for the remaining data and compare results with those obtained in the text for the entire data set.

8.7. REFERENCES

Federer, W. T. (1993). *Statistical Design and Analysis for Intercropping Experiments: Volume I: Two Crops*. Springer-Verlag, New York, Heidelberg, Berlin, Chapter 10, pp 1–298.

Federer, W. T. (1999). *Statistical Design and Analysis for Intercropping Experiments: Volume II: Three or More Crops*. Springer-Verlag, New York, Heidelberg, Berlin, pp 1–262.

Federer, W. T. (2005). Augmented split block experiment design. Agronomy J 97:578–586.

Federer, W. T. and D. Raghavarao. (1975). On augmented designs. Biometrics 31:29–35.

Federer W. T. and B. T. Scully. (1993). A parsimonious statistical design and breeding procedure for evaluating and selecting desirable characteristics over environments. Theoret App Genet 86:612–620.

Mejza, Iwona (1998). Characterization of certain split-block designs with a control. Biometric J 40(5):627–639.

APPENDIX 8.1. CODES FOR NUMERICAL EXAMPLE 8.1

The following SAS codes give the statistical analyses for PROC GLM and PROC MIXED for all 105 observations in the numerical example described above. The second part of the program gives the analysis for A controls and the sum of the new treatment effects with the B factor entries.

```
data asbed;
input Y R A B;
IF (A>9) THEN NEWA = 0; ELSE NEWA = 1;
IF (NEWA) THEN AN = 99; ELSE AN = A;
IF (B>6) THEN NEWB = 0; ELSE NEWB = 1;
IF (NEWB) THEN BN = 999; ELSE BN = B;
/*This part treats the new treatments as random effects.*/
/*The data set was not abbreviated as it was desired to show the
entries in the data file. R is replicate, A is factor A treatment, B is
factor B treatment, Y is response*/
datalines;

9   1   1   1
2   1   1   2
8   1   1   7
7   1   1   8
5   1   1   9
9   1   2   1
```

APPENDIX

```
7    1    2    2
3    1    2    7
5    1    2    8
4    1    2    9
9    1    3    1
2    1    3    2
8    1    3    7
7    1    3    8
5    1    3    9
9    1   10    1
1    1   10    2
9    1   10    7
7    1   10    8
5    1   10    9
9    1   11    1
7    1   11    2
3    1   11    7
5    1   11    8
4    1   11    9
9    1   12    1
2    1   12    2
8    1   12    7
7    1   12    8
5    1   12    9
9    1   13    1
7    1   13    2
3    1   13    7
5    1   13    8
4    1   13    9
9    2    4    3
7    2    4    4
13   2    4    7
8    2    4    8
8    2    4    9
9    2    5    3
12   2    5    4
8    2    5    7
7    2    5    8
8    2    5    9
9    2    6    3
7    2    6    4
13   2    6    7
9    2    6    8
12   2    6    9
9    2   10    3
11   2   10    4
9    2   10    7
7    2   10    8
5    2   10    9
```

9	2	11	3
7	2	11	4
13	2	11	7
5	2	11	8
4	2	11	9
9	2	12	3
12	2	12	4
8	2	12	7
7	2	12	8
5	2	12	9
9	2	13	3
7	2	13	4
13	2	13	7
5	2	13	8
4	2	13	9
19	3	7	5
17	3	7	6
13	3	7	7
15	3	7	8
14	3	7	9
19	3	8	5
12	3	8	6
18	3	8	7
17	3	8	8
45	3	8	9
19	3	9	5
17	3	9	6
13	3	9	7
25	3	9	8
34	3	9	9
15	3	10	5
9	3	10	6
11	3	10	7
10	3	10	8
10	3	10	9
9	3	11	5
17	3	11	6
13	3	11	7
15	3	11	8
14	3	11	9
9	3	12	5
12	3	12	6
8	3	12	7
17	3	12	8
15	3	12	9
9	3	13	5
17	3	13	6
13	3	13	7
15	3	13	8

```
14  3  13  9
;
/*ANOVA AND LSMEANS FOR DATA ASBED*/
PROC GLM DATA = asbed;
  CLASS R A B;
  MODEL Y = R A R*A B B*R A*B A*B*R;
  LSMEANS A B;
RUN;

/*ANOVA AND LSMEANS FOR CHECKS ONLY*/
PROC GLM DATA = work.asbed;
CLASS R A B AN BN;
MODEL Y = R AN AN*R BN BN*R AN*BN AN*BN*R;
LSMEANS AN BN AN*BN;
RUN;

/*This part of the code is for the analysis of all 105 observations.*/
PROC MIXED DATA = work.asbed;
  CLASS R A B BN AN ;
  MODEL Y = AN BN AN*BN/SOLUTION;
  RANDOM R A*NEWA B*NEWB/SOLUTION;
  LSMEANS AN BN AN*BN;
  MAKE 'SOLUTIONR' OUT = sr;
  RUN;
  PROC SORT DATA = sr;
  BY DESCENDING estimate;
RUN;
PROC PRINT; RUN;
```

To obtain the analysis for A checks and the B check treatments, use the following statement after the INPUT statement in PROC GLM:

IF B < 7 THEN DELETE;

The above statement deletes entries 1 to 6. The following statement is used to obtain the analysis of B checks and A check treatments in PROC GLM:

IF A < 10 THEN DELETE;

This statement deletes entries 1 to 9 of the A treatments. Following the above code, PROC MIXED codes could be included in the last two runs. Note that entries 99 and 999 are not included in the above two analyses, as they were in the first two after the data.

CHAPTER 9

Missing Observations in Split Plot and Split Block Experiment Designs

9.1. INTRODUCTION

Missing observations can occur as a result of many causes during the conduct of an experiment. Animals can invade and destroy some experimental units. Floods or fires can occur and damage a part of the experiment. On some occasions workers have been known to unintentionally leave out some of the experimental units when setting up the experiment. The problem of obtaining a statistical analysis of the results from an experiment with missing or damaged experimental units is resolvable. Several available statistical computer software packages handle this situation. Data analysis with missing observations is not more difficult than when there are no missing observations. SAS PROC GLM (SAS Institute, 1999–2001) is used for the numerical examples illustrating the data analysis with missing observations.

A split plot designed experiment with missing observations is presented in the next section. A split block designed experiment with missing observations is shown in Section 9.3. Whole plot, split plot, split block whole plot, or subplot experimental units can be missing. The SAS PROC GLM codes handle all these situations with a correct adjustment for the degrees of freedom associated with the missing observations in most cases. A discussion of missing observations for variations of the split plot and split block experiment designs is given. The SAS codes and outputs for the numerical examples are given in the two appendices, 9.1 and 9.2.

Variations on Split Plot and Split Block Experiment Designs, by Walter T. Federer and Freedom King
Copyright © 2007 John Wiley & Sons, Inc.

9.2. MISSING OBSERVATIONS IN A SPLIT PLOT EXPERIMENT DESIGN

To illustrate the computations for a split plot experiment design, we use a numerical example and the SAS PROC GLM procedure (SAS Institute, Inc., 1999-2001). Using the data of Example 1.1 in Chapter 1, omit two observations in replicate 4, planting method 4, and the responses for seedbed preparations 3 and 4, that is, data values 65.6 and 63.3. There are now 62 observations rather than 64. Using the SAS PROC GLM code as shown in Appendix 9.1, an analysis of variance table with Type III sums of squares is obtained and is presented below:

Source of variation	Degrees of freedom	Sum of squares	Mean square
Replicate = R	3	173.87	57.96
Seedbed preparation = A	3	214.02	71.34
Error A = A × R	9	97.38	10.82
Planting method = B	3	4100.79	1366.93
A × B	9	236.99	26.33
Error B = B × R/A	34	592.74	17.43

The Type I sums of squares and the estimable least squares means, lsmeans, are given in the output for Example 9.1 in Appendix 9.1. The F-tests proceed as for the equal numbers case, that is, the error term for factor A is Error A and the error term for factor B and the interaction of factors A and B is Error B. Several computer packages are able to handle this case where there are an unequal number of observations.

Instead of having missing observations in the split plot experimental units, whole plot experimental units may be missing. To illustrate this case, suppose that the disk-harrowed plots A4 in replicates 3 and 4 were missing. There would be two missing whole plots and eight missing split plot experimental units resulting in 56 data values. A partitioning of the degrees of freedom in an analysis table would be as follows:

Source of variation	Degrees of freedom
Total	56
Correction for the mean	1
Replicate = R	3
Seedbed preparation = A	3
Error A = A × R	7
Planting method = B	3
A × B	9
Error B = B × R/A	30

There were two missing whole plots and the two degrees of freedom for these are taken out of the Error A degrees of freedom. The $B \times R$ sum of squares within whole plots A has $9 + 9 + 9 + 3 = 30$ degrees of freedom. SAS PROC GLM provides the sums of squares and mean squares for the above partitioning of the degrees of freedom. The F-statistics may be obtained just as for no missing values.

9.3. MISSING OBSERVATIONS IN A SPLIT BLOCK EXPERIMENT DESIGN

Following the same steps as in the previous section, a numerical example is used to illustrate the computations for a split-block-designed experiment with missing observations. Using the data for Example 2.1 of Chapter 2, omit the last three observations for the example. These are for hybrid 10 in replicate 2 and are equal to 43, 43, and 42 for generations b, c, and a, respectively. The number of observations is reduced to 57 from 60, as present in the example in Chapter 2. From the computer output for Example 9.2, Appendix 9.2, the following Type III analysis of variance is obtained:

Source of variation	Degrees of freedom	Sum of squares	Mean square
Replicate = R	1	0.17	0.17
Hybrid = H	9	66.80	7.42
Error $H = H \times R$	8	67.00	8.38
Generation = G	2	30.67	15.34
Error $G = G \times R$	2	12.11	6.06
$G \times H$	18	60.50	3.36
Error $GH = G \times H \times R$	16	22.22	1.39

One degree of freedom is lost from Error A and two from the three factor interaction $G \times H \times R$. As may be seen when using available software, missing observations present no difficulties in analyzing data.

9.4. COMMENTS

As demonstrated in Chapters 3, 4, 5, and 6, there are many variations of split plot and split block experiment designs. When missing observations occur, use of the same computer codes as for no missing observations provide the statistical analysis in the same forms. Since orthogonality is disturbed by the missing observations, Type III or Type IV analyses should be used. Statistical analyses without the use of computer software can become cumbersome. A note of caution in using software packages is to always check on the number of degrees of freedom to ascertain that they are correct. F-test statistics may be computed as described previously.

9.5. PROBLEMS

Problem 9.1. Omit another observation (e.g., replicate 1, planting method B1, and seedbed preparation A1) for the example discussed in Section 9.2 and perform an analysis of the remaining data.

PROBLEMS

Problem 9.2. Omit another observation (e.g., replicate 1, hybrid 7, and generation a) for the example discussed in Section 9.3 and perform an analysis of the remaining data.

Problem 9.3. A study was conducted to investigate the effect of gender (whole plot factor), age group (split plot factor), dieting (split split plot factor), and exercise regimen (split split split plot factor) on weight loss. A random sample of 300 males and 300 females was selected as the experimental subjects. The number of subjects per combination of the four factors is presented in the table that follows. Note that the numbers are unequal.

Age group	Diet	Exercise	Female	Male
Young	No	0	8	10
Young	No	1	8	10
Young	No	2	8	10
Young	No	3	8	10
Young	No	4	8	10
Young	Yes	0	8	10
Young	Yes	1	8	10
Young	Yes	2	8	10
Young	Yes	3	8	10
Young	Yes	4	8	10
Middle age	No	0	12	12
Middle age	No	1	12	12
Middle age	No	2	12	12
Middle age	No	3	12	12
Middle age	No	4	12	12
Middle age	Yes	0	12	12
Middle age	Yes	1	12	12
Middle age	Yes	2	12	12
Middle age	Yes	3	12	12
Middle age	Yes	4	12	12
Old	No	0	10	8
Old	No	1	10	8
Old	No	2	10	8
Old	No	3	10	8
Old	No	4	10	8
Old	Yes	0	10	8
Old	Yes	1	10	8
Old	Yes	2	10	8
Old	Yes	3	10	8
Old	Yes	4	10	8
Total			300	300

(i) Obtain a partitioning of the 600 degrees of freedom into the degrees of freedom for each source of variation in an analysis of variance table.

(ii) Write a SAS PROC GLM code for obtaining an analysis of variance table, F-tests, and means for all combinations and their standard errors.

(iii) Are Type I sums of squares equal to Type III sums of squares? Why or why not?

(iv) Simulate 600 numbers using random normal deviates plus 5 to form a data set and use your code to analyze the data set.

9.6. REFERENCE

SAS Institute, Inc. (1999–2001). Release 8.02, copyright. Cary, NC.

APPENDIX 9.1. SAS CODE FOR NUMERICAL EXAMPLE IN SECTION 9.2.

A computer code and data for the numerical example in Section 9.2 is given below:

```
Data spex1;
input Y R A B; /*Y=yield, R=block, A=planting method, B=cultivation method*/datalines;

82.8   1   1   1
46.2   1   1   2
78.6   1   1   3
77.7   1   1   4
72.2   2   1   1
51.6   2   1   2
70.9   2   1   3
73.6   2   1   4
72.9   3   1   1
53.6   3   1   2
69.8   3   1   3
70.3   3   1   4
74.6   4   1   1
57.0   4   1   2
69.6   4   1   3
72.3   4   1   4
74.1   1   2   1
49.1   1   2   2
72.0   1   2   3
66.1   1   2   4
76.2   2   2   1
53.8   2   2   2
71.8   2   2   3
65.5   2   2   4
```

APPENDIX

```
71.1  3  2  1
43.7  3  2  2
67.6  3  2  3
66.2  3  2  4
67.8  4  2  1
58.8  4  2  2
60.6  4  2  3
60.6  4  2  4
68.4  1  3  1
54.5  1  3  2
72.0  1  3  3
70.6  1  3  4
68.2  2  3  1
47.6  2  3  2
76.7  2  3  3
75.4  2  3  4
67.1  3  3  1
46.4  3  3  2
70.7  3  3  3
66.2  3  3  4
65.6  4  3  1
53.3  4  3  2
65.6  4  3  3
69.2  4  3  4
71.5  1  4  1
50.9  1  4  2
76.4  1  4  3
75.1  1  4  4
70.4  2  4  1
65.0  2  4  2
75.8  2  4  3
75.8  2  4  4
72.5  3  4  1
54.9  3  4  2
67.6  3  4  3
75.2  3  4  4
67.8  4  4  1
50.2  4  4  2   /*last 2 observations of Example 1.1 omitted*/

;


Proc GLM;

Class R A B;
Model Y = R A R*A B A*B;
Lsmeans A B A*B;
Run;
```

The computer output from the above code and data set is presented below:

The GLM Procedure

Dependent Variable: Y

Source	DF	Sum of Squares	Mean Square	F Value	Pr > F
Model	27	4945.101519	183.151908	10.51	<.0001
Error	34	592.735417	17.433395		
Corrected Total	61	5537.836935			

R-Square	Coeff Var	Root MSE	Y Mean
0.892966	6.305765	4.175332	66.21452

Source	DF	Type I SS	Mean Square	F Value	Pr > F
R	3	221.932918	73.977639	4.24	0.0119
A	3	199.166118	66.388706	3.81	0.0187
R*A	9	189.280400	21.031156	1.21	0.3232
B	3	4097.733083	1365.911028	78.35	<0001
A*B	9	236.989000	26.332111	1.51	0.1840

Source	DF	Type III SS	Mean Square	F Value	Pr > F
R	3	173.866083	57.955361	3.32	0.0311
A	3	214.023083	71.341028	4.09	0.0139
R*A	9	97.381190	10.820132	0.62	0.7710
B	3	4100.789391	1366.929797	78.41	<0001
A*B	9	236.989000	26.332111	1.51	0.1840

Least Squares Means

A	Y LSMEAN
1	68.3562500
2	64.0625000
3	64.8437500
4	67.9583333

B	Y LSMEAN
1	71.4500000
2	52.2875000
3	70.8604167
4	70.6229167

A	B	Y LSMEAN
1	1	75.6250000
1	2	52.1000000
1	3	72.2250000
1	4	73.4750000

2	1	72.3000000
2	2	51.3500000
2	3	68.0000000
2	4	64.6000000
3	1	67.3250000
3	2	50.4500000
3	3	71.2500000
3	4	70.3500000
4	1	70.5500000
4	2	55.2500000
4	3	71.9666667
4	4	74.0666667

APPENDIX 9.2. SAS CODE FOR NUMERICAL EXAMPLE IN SECTION 9.3.

The computer code for the data of the example in Section 9.3 is presented below:

```
data sbex;
input yield rep hyb gen;
datalines;

48 1 3 1
46 1 3 3
43 1 3 2
46 1 8 1
45 1 8 3
42 1 8 2
46 1 2 1
44 1 2 3
42 1 2 2
42 1 1 1
46 1 1 3
44 1 1 2
43 1 6 1
45 1 6 3
44 1 6 2
47 1 7 1
49 1 7 3
47 1 7 2
48 1 0 1
45 1 0 3
45 1 0 2
46 1 9 1
48 1 9 3
47 1 9 2
46 1 4 1
48 1 4 3
```

47 1 4 2
49 1 5 1
49 1 5 3
48 1 5 2
46 2 4 2
48 2 4 3
42 2 4 1
45 2 3 2
44 2 3 3
42 2 3 1
46 2 9 2
46 2 9 3
44 2 9 1
45 2 5 2
45 2 5 3
43 2 5 1
43 2 1 2
50 2 1 3
44 2 1 1
48 2 7 2
51 2 7 3
48 2 7 1
44 2 2 2
48 2 2 3
47 2 2 1
44 2 8 2
46 2 8 3
46 2 8 1
47 2 6 2
48 2 6 3
44 2 6 1 /*last 3 observations for Example 2.1 were omitted for this example.*/
;

```
proc glm data = sbex;
  class rep hyb gen;
  model yield = rep hyb hyb*rep gen gen*rep gen*hyb;
  lsmeans hyb gen gen*hyb;
run;
```

The output of the above code and data set is presented below in an abbreviated form:

```
              Class Level Information
       Class      Levels     Values

       rep          2        1 2
       hyb         10        0 1 2 3 4 5 6 7 8 9
       gen          3        1 2 3
```

APPENDIX

Number of observations   57
Dependent Variable: yield

Source	DF	Sum of Squares	Mean Square	F Value	Pr > F
Model	40	247.8128655	6.1953216	4.46	0.0011
Error	16	22.2222222	1.3888889		
Corrected Total	56	270.0350877			

R-Square	Coeff Var	Root MSE	yield Mean
0.917706	2.574747	1.178511	45.77193

Source	DF	Type I SS	Mean Square	F Value	Pr > F
rep	1	0.23879142	0.23879142	0.17	0.6839
hyb	9	66.79629630	7.42181070	5.34	0.0018
rep*hyb	8	67.00000000	8.37500000	6.03	0.0012
gen	2	36.35087719	18.17543860	13.09	0.0004
rep*gen	2	16.92319688	8.46159844	6.09	0.0108
hyb*gen	18	60.50370370	3.36131687	2.42	0.0409

Source	DF	Type III SS	Mean Square	F Value	Pr > F
rep	1	0.16666667	0.16666667	0.12	0.7335
hyb	9	66.79629630	7.42181070	5.34	0.0018
rep*hyb	8	67.00000000	8.37500000	6.03	0.0012
gen	2	30.67111111	15.33555556	11.04	0.0010
rep*gen	2	12.11111111	6.05555556	4.36	0.0308
hyb*gen	18	60.50370370	3.36131687	2.42	0.0409

Least Squares Means

hyb	yield LSMEAN
0	Non-est
1	44.8333333
2	45.1666667
3	44.6666667
4	46.1666667
5	46.5000000
6	45.1666667
7	48.3333333
8	44.8333333
9	46.1666667

gen	yield LSMEAN
1	Non-est
2	Non-est
3	Non-est

hyb	gen	yield LSMEAN
0	1	Non-est
0	2	Non-est
0	3	Non-est

1	1	43.0000000
1	2	43.5000000
1	3	48.0000000
2	1	46.5000000
2	2	43.0000000
2	3	46.0000000
3	1	45.0000000
3	2	44.0000000
3	3	45.0000000
4	1	44.0000000
4	2	46.5000000
4	3	48.0000000
5	1	46.0000000
5	2	46.5000000
5	3	47.0000000
6	1	43.5000000
6	2	45.5000000
6	3	46.5000000
7	1	47.5000000
7	2	47.5000000
7	3	50.0000000
8	1	46.0000000
8	2	43.0000000
8	3	45.5000000
9	1	45.0000000
9	2	46.5000000
9	3	47.0000000

CHAPTER 10

Combining Split Plot or Split Block Designed Experiments over Sites

10.1. INTRODUCTION

Explicit procedures for combining results from a group of split plot designed experiments and a group of split block designed experiments are given in Sections 10.2 and 10.3. Numerical examples are used to illustrate the use of SAS PROC GLM and SAS PROC MIXED codes in obtaining the analyses for the data. A Tukey studentized range multiple comparisons procedure is also illustrated. As will be seen, the procedures given are straightforward and do not increase the difficulty to any extent. Some comments about combining other types of split plot and/or split block designed experiments are presented in Section 10.4.

10.2. COMBINING SPLIT PLOT DESIGNED EXPERIMENTS OVER SITES

When a standard split plot designed experiment is conducted at several sites, in several years, or repeated in some other fashion, the researcher may want to combine the results from the individual experiments. The results of the individual experiments will need to be obtained and interpreted as well. Suppose that a standard split plot design as described in Chapter 1 is used at s different sites. There are r replicates, a whole plot treatments, factor A, and b split plot treatments, factor B, at each site. The analyses of variance tables at each site are of the following form

---

Variations on Split Plot and Split Block Experiment Designs, by Walter T. Federer and Freedom King
Copyright © 2007 John Wiley & Sons, Inc.

where the linear model at each site is that for the standard split plot experiment design:

Source of variation	Degrees of freedom	Sum of squares at each site			
		Site 1	Site 2	Site 3 ...	Site s
Total	rab	T1	T2	T3	Ts
Correction for mean	1	C1	C2	C3	Cs
Replicate, R	$r-1$	R1	R2	R3	Rs
Factor A	$a-1$	A1	A2	A3	As
Error $= R \times A$	$(r-1)(a-1)$	EA1	EA2	EA3	EAs
Factor B	$b-1$	B1	B2	B3	Bs
$A \times B$	$(a-1)(b-1)$	AB1	AB2	AB3	ABs
Error $B = R \times B/A$	$a(b-1)(r-1)$	EB1	EB2	EB3	EBs

The sums of squares for the analysis of variance at any site are the usual ones for a standard split plot experiment design. From the above, it is noted that the sum of squares for the Error $B = R \times B/A$, $R \times B$ within levels of factor A, is the $R \times B$ sum of squares summed over the a whole plot treatments, factor A. In this form, it becomes obvious what the "Error B" sum of squares is. Likewise, as presented in the above form, the individual experiment results can be studied and interpreted at each site.

A linear model for combining split plot experiment designs over sites is obtained from the sources of variation given in the following ANOVA table or from the SAS PROC GLM code in Appendix 10.1. A combined partitioning of the degrees of freedom and sums of squares in an analysis of variance table for the s experiments may be of the following form:

Source of variation	Degrees of freedom	Sum of squares
Total	$srab$	$\sum_{i=1}^{s} T_i$
Correction terms within sites	s	$\sum_{i=1}^{s} C_i$
Overall correction term	1	G
Sites, S	$s-1$	$\sum_{i=1}^{s} C_i - G$
R within sites	$s(r-1)$	$\sum_{i=1}^{s} R_i$
A within sites	$s(a-1)$	$\sum_{i=1}^{s} A_i$
A	$a-1$	Compute
$A \times S$	$(a-1)(s-1)$	by subtraction
EA within sites	$s(a-1)(r-1)$	$\sum_{i=1}^{s} EA_i$
B within sites	$s(b-1)$	$\sum_{i=1}^{s} B_i$
B	$b-1$	Compute
$B \times S$	$(b-1)(s-1)$	by subtraction
$A \times B$ within sites	$s(a-1)(b-1)$	$\sum_{i=1}^{s} AB_i$
$A \times B$	$(a-1)(b-1)$	Compute
$A \times B \times S$	$(a-1)(b-1)(s-1)$	by subtraction
EB within sites	$sa(b-1)(r-1)$	$\sum_{i=1}^{s} EB_i$

A SAS PROC GLM MODEL statement for obtaining the above analysis of variance table would be:

$$Y = S\ R(S)\ A\ A^*S\ A^*R(S)\ B\ B^*S\ A^*B\ A^*B^*S;$$

One should not partition the replicate within site sum of squares into a replicate sum of squares and a replicate by site interaction. Doing this is incorrect owing to the fact that the numbering of the replicates at each site is arbitrary. Replicate 1 at site 1 has nothing to do with replicate 1 at site 2, as this replicate could have been numbered 3 or any other number. The replicates numbered 1, for example, have nothing in common, except the number 1. There is no such item as a replicate 1 effect over sites. The replicate effect is a nested within site effect.

The above configuration of an analysis of variance table helps to clarify what the various within site sums of squares contain. A two-factor interaction sums of squares, $A \times B$, nested within sites contains the two-factor interaction $A \times B$ and the three-factor interaction $A \times B \times S$.

Several items need to be investigated when combining results of experiments over sites. The Error B variances may vary from site to site resulting in variance heterogeneity. This would be a form of the Behrens-Fisher situation. The procedures presented by Grimes and Federer (1984) may be used when variance heterogeneity is present. Alternatively, one may use a variance stabilizing transformation of the data such as logarithms or square roots. Another transformation is to obtain the $A \times B$ interaction means, to divide the *ab* means by the standard error of a mean at site i, and to run the analysis on the *abs* transformed means for the three factors A, B, and S. This transformation tends to make the means have a unit normal distribution with variance equal to one. A partitioning of the degrees of freedom for the last transformation in an analysis of variance table is presented below:

Source of variation	Degrees of freedom
Sites, S	$s - 1$
Factor A	$a - 1$
$A \times S$	$(a-1)(s-1)$
Factor B	$b - 1$
$A \times B$	$(a-1)(b-1)$
$B \times S$	$(b-1)(s-1)$
$A \times B \times S$	$(a-1)(b-1)(s-1)$
Error	infinite as the variance is known to be one

This transformation is a standardization of the data and was one of the procedures used by Federer et al. (2001) to combine results from experiments with different experiment designs and unequal error variances at the different sites.

Example 10.1—A numerical example illustrating the analysis for a split plot designed experiment repeated over s sites is described here. Suppose the whole plot treatments are a factor A by factor B factorial with $a = 5$ factor A levels and with $b = 2$ factor B levels arranged in a randomized complete block experiment design with $r = 3$ replicates (blocks). Furthermore, suppose the number of factor C split plot treatments is $c = 4$ and the number of sites is $s = 3$. A simulated data set of the 480 observations for the design described here is given in Appendix 10.1. Since the number of combinations is large, an analyst might wish to use multiple comparisons such as Tukey's studentized range test. This procedure is demonstrated for the SAS PROC GLM for factor A, B, and C means. The GLM code in written form does not allow multiple comparisons procedures of means for combinations of factors. However, SAS PROC MIXED does provide such pairwise comparisons, as demonstrated with the output given in Appendix 10.1.

The factor site was considered to be a fixed effect as some experimenters often select a group of sites with the desired characteristics and use these sites for comparing various other factors. This puts the factor site in the fixed effect category. This is the manner used to write the codes. If the analyst desires to consider site as a random effect, the codes should be altered accordingly.

An analysis of variance table for the data from this example is given in Table 10.1. Note that this is an orthogonal arrangement for all factors and hence the Type I and III analyses are identical.

The different F-statistics and the probability of obtaining a larger value of F were obtained using the code in Appendix 10.1. Note that the degrees of freedom for the $AB \times R(S)$ sum of squares is $s(ab - 1)(r - 1) = 4\{[5(2) - 1][3 - 1]\} = 72$. SAS uses the notation $A*B*R(S)$ to obtain this term.

Table 10.1. Analysis of Variance for the Data of Example 10.1 with Associated F-Tests.

Source of variation	Degrees of freedom	Sum of squares	Mean square	F-value	$P > F$
Total (corr.)	479	1640323006	—	—	—
Site, S	3	552717	184239	0.21	0.8876
Block(site), R(S)	8	7062320	882790	—	—
A	4	1387680917	346920229	133637	<.0001
A × S	12	34068	2839	1.09	0.3787
B	1	100939695	100939695	38883	<.0001
B × S	3	1618	539	0.21	0.8908
A × B	4	31444008	7861002	3028	<.0001
A × B × S	12	33737	2811	1.08	0.3872
AB × R(S)	72	186911	2596	—	—
C	3	19356264	6452088	2033.43	<.0001
A × C	12	26075792	2172983	684.83	<.0001
B × C	3	23901388	7967129	2510.91	<.0001
A × B × C	12	41996729	3499727	1102.97	<.0001
C × S	9	47625	5292	1.67	0.0975
A × C × S	36	104110	2892	0.91	0.6177
B × C × S	9	61111	6790	2.14	0.0270
A × B × C × S	36	82475	2291	0.72	0.8792
Error C	240	761522	3173	—	—

10.3. COMBINING SPLIT BLOCK DESIGNED EXPERIMENTS OVER SITES

A similar procedure as described in the previous section is used for combining the results from split block designed experiments over several sites or conditions. Suppose that there are s sites over which a split block designed experiment with r replicates, a levels of factor A, and b levels of factor B, has been conducted. The linear model used for each site is that for a standard split block experiment design as given in Chapter 2. A partitioning of the degrees of freedom and sums of squares in analysis of variance tables at each of the s sites is given below:

Source of variation	Degrees of freedom	Sums of squares at each site				
		1	2	3	...	s
Total	rab	T1	T2	T3		Ts
Correction for mean	1	C1	C2	C3		Cs
Replicate = R	$r-1$	R1	R2	R3		Rs
Factor A	$a-1$	A1	A2	A3		As
Error $A = A \times R$	$(a-1)(r-1)$	EA1	EA2	EA3		EAs
Factor B	$b-1$	B1	B2	B3		Bs
Error $B = B \times R$	$(b-1)(r-1)$	EB1	EB2	EB3		EBs
$A \times B$	$(a-1)(b-1)$	AB1	AB2	AB3		ABs
Error $AB = A \times B \times R$	$(a-1)(b-1)(r-1)$	EAB1	EAB2	EAB3		EABs

In this format, the mean squares may be obtained and the results for each site may be studied and interpreted for each of the individual experiments.

A linear model for the analysis of data from experiments designed as a standard split block experiment design at s sites, is given by the sources of variation in the following ANOVA table. A combined analysis of variance table would have the following partitioning of the $rabs$ degrees of freedom and sums of squares:

Source of variation	Degrees of freedom	Sum of squares
Total	$rabs$	$\sum_{i=1}^{s} T_i$
Correction for mean within sites	s	$\sum_{i=1}^{s} C_i$
Correction for mean	1	compute
Sites = S	$s-1$	by subtraction
Replicate within sites	$s(r-1)$	$\sum_{i=1}^{s} R_i$
Factor A within sites	$s(a-1)$	$\sum_{i=1}^{s} A_i$
Factor A	$a-1$	compute
$A \times S$	$(a-1)(s-1)$	by subtraction
$A \times R$ within sites	$s(a-1)(r-1)$	$\sum_{i=1}^{s} EA_i$
Factor B within sites	$s(b-1)$	$\sum_{i=1}^{s} B_i$

(Continued)

(Continued)

Source of variation	Degrees of freedom	Sum of squares
B	$b-1$	compute
$B \times S$	$(b-1)(s-1)$	by subtraction
$B \times R$ within sites	$s(b-1)(r-1)$	$\sum_{i=1}^{s} EB_i$
$A \times B$ within sites	$s(a-1)(b-1)$	$\sum_{i=1}^{s} AB_i$
$A \times B$	$(a-1)(b-1)$	compute
$A \times B \times S$	$(a-1)(b-1)(s-1)$	by subtraction
$A \times B \times R$ within sites	$s(a-1)(b-1)(r-1)$	$\sum_{i=1}^{s} EAB_i$

The same comments about variance heterogeneity and transformation of data given in the previous section apply here as well. F-tests are carried out as usual. For example the $A \times R$ within site mean square, is used as the error mean square for testing hypotheses about factor A effects and the $A \times S$ interaction effects.

Computer software such as SAS is available for obtaining the above analysis of variance table and for computing F-statistics. The statements for computing F-statistics in the SAS code were described in Chapter 1 and are obtained with the code presented in Appendix 10.2. A SAS PROC GLM MODEL statement for the above analysis is

$$Y = S\ R(S)\ A\ A^*S\ A^*R(S)\ B\ B^*S\ B^*R(S)\ A^*B\ A^*B^*S;$$

Example 10.2—To illustrate the statistical analysis for a split block designed experiment repeated over sites, data were simulated and appear in Appendix 10.2. For the split block experiment design, let the number of factor A levels be $a = 5$ and the number of factor B levels be $b = 8$ with each factor arranged in a randomized complete block experiment design with $r = 4$ replicates. Furthermore, suppose this experiment is repeated over $s = 3$ sites. An analysis of variance table with associated F-statistics is given in Table 10.2. Note that this is an orthogonal arrangement and Type I and Type III sums of squares are identical.

Table 10.2. Analysis of Variance and Associated F-Tests for the Data of Example 10.2.

Source of variation	Degrees of freedom	Sum of squares	Mean square	F-value	$P > F$
Total (corr.)	479	6371649132	—	—	—
Site, S	2	523573968	261786984	0.63	0.5559
Block (S), $R(S)$	9	3756646710	417405190	—	—
A	4	29288163	7322041	147.80	<.0001
$A \times S$	8	247899	30987	0.63	0.7508
$A \times R(S)$	36	1783391	49539	—	—
B	7	1937592291	276798899	164.94	<.0001
$B \times S$	14	15903698	1135978	0.68	0.7878
$B \times R(S)$	63	105727288	1678211	—	—
$A \times B$	28	91141	3255	1.25	0.1838
$A \times B \times S$	56	140534	2510	0.97	0.5461
Error	252	654049	2595	—	—

10.4. DISCUSSION

The same procedures as described above apply equally well to sets of any of the variations of split plot experiment designs and split block experiment designs. All of the results from a group of experiments such as described above may be entered in a single data file. Statistical analyses for each of the experiments may be obtained by using SAS IF-THEN statements. Or for example, suppose the analysis at site three is desired. To do this, use the following statements after the INPUT statement:

```
IF S < 3 THEN OMIT;
IF S > 3 THEN OMIT;
```

10.5. PROBLEMS

Problem 10.1. Suppose the standard split plot design of Example 1.1 was repeated over $s = 3$ sites. Simulate similar data for the two additional sites and obtain the analysis over the three sites as described above.

Problem 10.2. Delete site 1 data and obtain an analysis of the data for the remaining data of Example 10.1. Obtain the Tukey studentized range test for the A*B combinations.

Problem 10.3. Using the data of Example 2.1, simulate three additional data sets and obtain the combined analysis of data for the four data sets (sites).

Problem 10.4. Delete the data for site 3 and obtain an analysis for the remaining data in Example 10.2. Perform the lsd multiple comparisons procedure for the A*B*S means.

10.6. REFERENCES

Federer, W. T., M. Reynolds, and J. Crossa. (2001). Combining results of augmented designs over sites. Agronomy Journal 93:389–395.

Grimes, B. A. and W. T. Federer. (1984). Comparison of means from populations with unequal variances. In *W. G. Cochran's Impact on Statistics*, P. S. R. S. Rao and J. Sedransk, Editors), John Wiley & Sons, Inc., New York, pp. 353–374.

APPENDIX 10.1. EXAMPLE 10.1

The data set for this example is presented on the book's FTP site. The SAS GLM and MIXED codes and a subset of the data are presented below.

```
/*Simulated data: split plot across sites
whole plot factors: A and B are in a randomized complete block design
split plot factor: C
```

number of sites = 4
number of blocks at each site = 3*/

```
options ls = 79 nocenter nodate pageno = 1;
data New;
input Obs Site Block $ A $ B $ C $ Yield;
datalines;
    1   1    R1   A1   B1   C1   6979
    2   1    R1   A1   B1   C2   7272
    3   1    R1   A1   B1   C3   7565
    4   1    R1   A1   B1   C4   7827
    5   1    R1   A1   B2   C1   8113
    6   1    R1   A1   B2   C2   7025
    7   1    R1   A1   B2   C3   7340
    8   1    R1   A1   B2   C4   7637
    9   1    R1   A2   B1   C1   7910
    .
    .
    .
  476   4    R3   A5   B1   C4  12300
  477   4    R3   A5   B2   C1  12500
  478   4    R3   A5   B2   C2  12900
  479   4    R3   A5   B2   C3  13000
  480   4    R3   A5   B2   C4  13700

proc glm data = new;
class Site Block C B A;
model Yield = Site Block(Site)A A*site B B*site A*B A*B*site
A*B*Block(site) C
A*C B*C A*B*C C*site A*C*site B*C*site A*B*C*site/ss3;
random Site Block(site) A*B*Block(site)/test;
lsmeans A B C;
/*lsmeans A*site;              /*Means not computed*/
lsmeans A*B A*C B*C A*B*C;
lsmeans A*site B*site A*B*site;
lsmeans C*site A*C*site B*C*site A*B*C*site; */

/*multiple comparisons using Tukey*/

means A B /tukey alpha = 0.05 e = A*B*Block(site);
means C /tukey alpha = 0.05;
run;

/* Using proc mixed for multiple comparisons using Tukey*/

proc mixed data = new;
class Site A B C Block;
model Yield = site A A*site B B*site A*B A*B*site C A*C B*C A*B*C
C*site A*C*site B*C*site A*B*C*site;
random Block(Site) A*B*Block(Site);
```

APPENDIX

```
* LS means per factor;

lsmeans A B;
lsmeans A*site;
/*lsmeans A*B;   /*These means are of use to the researcher,
lsmeans B*site;  but were not computed*/
lsmeans C;
lsmeans A*B;
lsmeans A*B*site;
lsmeans B*C;
lsmeans A*C;
lsmeans A*B*C;*/
lsmeans A/adjust = tukey alpha = 0.05;
lsmeans B/adjust = tukey alpha = 0.05;
lsmeans A*site/adjust = tukey alpha = 0.05;
run;
```

The output for the above code and data set, in a drastically abbreviated form, is presented below

The SAS System   1
The GLM Procedure

           Class Level Information
Class   Levels   Values
Site   4   1 2 3 4
Block   3   R1 R2 R3
C   4   C1 C2 C3 C4
B   2   B1 B2
A   5   A1 A2 A3 A4 A5
Number of observations   480

Dependent Variable: Yield
                       Sum of
Source   DF   Squares   Mean Square   F Value   Pr > F
Model   239 1639561484   6860090   2162.01   <.0001
Error   240 761522   3173
Corrected Total   479 1640323006

R-Square   Coeff Var   Root MSE   Yield Mean
0.999536   0.565140   56.32947   9967.354

Source   DF   Type III SS   Mean Square   F Value   Pr > F
Site   3   552717   184239   58.06   <.0001
Block(Site)   8   7062320   882790   278.22   <.0001
A   4   1387680917   346920229   109335   <.0001
Site*A   12   34068   2839   0.89   0.5530
B   1   100939695   100939695   31812.0   <.0001
Site*B   3   1618   539   0.17   0.9166
B*A   4   31444008   7861002   2477.46   <.0001
Site*B*A   12   33737   2811   0.89   0.5618
Block*B*A(Site)   72   186911   2596   0.82   0.8415

C	3	19356264	6452088	2033.43	<.0001
C*A	12	26075792	2172983	684.83	<.0001
C*B	3	23901388	7967129	2510.91	<.0001
C*B*A	12	41996729	3499727	1102.97	<.0001
Site*C	9	47625	5292	1.67	0.0975
Site*C*A	36	104110	2892	0.91	0.6177
Site*C*B	9	61111	6790	2.14	0.0270
Site*C*B*A	36	82475	2291	0.72	0.8794

The GLM Procedure

Source	Type III Expected Mean Square
Site	Var(Error) + 4 Var(Block*B*A(Site)) + 40 Var(Block(Site)) + 120 Var(Site) + Q(Site*A,Site*B,Site*B*A,Site*C,Site*C*A,Site*C*B,Site*C*B*A)
Block(Site)	Var(Error) + 4 Var(Block*B*A(Site)) + 40 Var(Block(Site))
A	Var(Error) + 4 Var(Block*B*A(Site)) + Q(A,Site*A,B*A,Site*B*A,C*A,C*B*A,Site*C*A,Site*C*B*A)
Site*A	Var(Error) + 4 Var(Block*B*A(Site)) + Q(Site*A,Site*B*A,Site*C*A,Site*C*B*A)
B	Var(Error) + 4 Var(Block*B*A(Site)) + Q(B,Site*B,B*A,Site*B*A,C*B,C*B*A,Site*C*B,Site*C*B*A)
Site*B	Var(Error) + 4 Var(Block*B*A(Site)) + Q(Site*B,Site*B*A,Site*C*B,Site*C*B*A)
B*A	Var(Error) + 4 Var(Block*B*A(Site)) + Q(B*A,Site*B*A,C*B*A,Site*C*B*A)
Site*B*A	Var(Error) + 4 Var(Block*B*A(Site)) + Q(Site*B*A,Site*C*B*A)
Block*B*A(Site)	Var(Error) + 4 Var(Block*B*A(Site))
C	Var(Error) + Q(C,C*A,C*B,C*B*A,Site*C,Site*C*A,Site*C*B,Site*C*B*A)
C*A	Var(Error) + Q(C*A,C*B*A,Site*C*A,Site*C*B*A)
C*B	Var(Error) + Q(C*B,C*B*A,Site*C*B,Site*C*B*A)
C*B*A	Var(Error) + Q(C*B*A,Site*C*B*A)
Site*C	Var(Error) + Q(Site*C,Site*C*A,Site*C*B,Site*C*B*A)
Site*C*A	Var(Error) + Q(Site*C*A,Site*C*B*A)
Site*C*B	Var(Error) + Q(Site*C*B,Site*C*B*A)
Site*C*B*A	Var(Error) + Q(Site*C*B*A)

The GLM Procedure
Tests of Hypotheses for Mixed Model Analysis of Variance
Dependent Variable: Yield

Source	DF	Type III SS	Mean Square	F Value	Pr > F
* Site	3	552717	184239	0.21	0.8876
Error	8	7062320	882790		

Error: MS(Block(Site))
* This test assumes one or more other fixed effects are zero.

APPENDIX

Source	DF	Type III SS	Mean Square	F Value	Pr > F
Block(Site)	8	7062320	882790	340.06	<.0001
* A	4	1387680917	346920229	133637	<.0001
* Site*A	12	34068	2838.991319	1.09	0.3787
* B	1	100939695	100939695	38882.9	<.0001
* Site*B	3	1617.616667	539.205556	0.21	0.8908
* B*A	4	31444008	7861002	3028.13	<.0001
* Site*B*A	12	33737	2811.432986	1.08	0.3872
Error	72	186911	2595.991667		

Error: MS(Block*B*A(Site))
* This test assumes one or more other fixed effects are zero.

Source	DF	Type III SS	Mean Square	F Value	Pr > F
Block*B*A(Site)	72	186911	2595.991667	0.82	0.8415
* C	3	19356264	6452088	2033.43	<.0001
* C*A	12	26075792	2172983	684.83	<.0001
* C*B	3	23901388	7967129	2510.91	<.0001
* C*B*A	12	41996729	3499727	1102.97	<.0001
* Site*C	9	47625	5291.691667	1.67	0.0975
* Site*C*A	36	104110	2891.932986	0.91	0.6177
* Site*C*B	9	61111	6790.146296	2.14	0.0270
Site*C*B*A	36	82475	2290.966319	0.72	0.8794
Error: MS(Error)	240	761522	3173.009028		

* This test assumes one or more other fixed effects are zero.

Least Squares Means

A	Yield LSMEAN
A1	7498.9792
A2	8810.6563
A3	10017.3854
A4	11193.0833
A5	12316.6667

B	Yield LSMEAN
B1	9508.7792
B2	10425.9292

C	Yield LSMEAN
C1	9706.2667
C2	9874.5500
C3	10039.6500
C4	10248.9500

Tukey's Studentized Range (HSD) Test for Yield
NOTE: This test controls the Type I experimentwise error rate, but it generally has a higher Type II error rate than REGWQ.

Alpha	0.05
Error Degrees of Freedom	72
Error Mean Square	2595.992
Critical Value of Studentized Range	3.95712

Minimum Significant Difference   20.578

Means with the same letter are not significantly different.

T
u
k
.
.
.
n
g   Mean   N   A
A   12316.667   96   A5
B   11193.083   96   A4
C   10017.385   96   A3
D   8810.656   96   A2
E   7498.979   96   A1

Tukey's Studentized Range (HSD) Test for Yield
NOTE: This test controls the Type I experimentwise error rate, but it generally has a higher Type II error rate than REGWQ.

Alpha   0.05
Error Degrees of Freedom   72
Error Mean Square   2595.992
Critical Value of Studentized Range   2.81929
Minimum Significant Difference   9.2723

Means with the same letter are not significantly different.

T
u
k
.
.
.
g   Mean   N   B
A   10425.929   240   B2
B   9508.779   240   B1

Tukey's Studentized Range (HSD) Test for Yield
NOTE: This test controls the Type I experimentwise error rate, but it generally has a higher Type II error rate than REGWQ.

Alpha   0.05
Error Degrees of Freedom   240
Error Mean Square   3173.009
Critical Value of Studentized Range   3.65877
Minimum Significant Difference   18.814

Means with the same letter are not significantly different.

T
u
k

```
n
g      Mean       N    C
A   10248.950   120   C4
B   10039.650   120   C3
C    9874.550   120   C2
D    9706.267   120   C1
```

The Mixed Procedure
      Class Level Information

```
Class    Levels   Values
Site        4     1 2 3 4
A           5     A1 A2 A3 A4 A5
B           2     B1 B2
C           4     C1 C2 C3 C4
Block       3     R1 R2 R3

Total Observations         480
```

The Mixed Procedure

```
  Covariance Parameter
        Estimates
Cov Parm          Estimate
Block(Site)         21993
A*B*Block(Site)         0
Residual          3039.85
```

Type 3 Tests of Fixed Effects

Effect	Num DF	Den DF	F Value	Pr > F
Site	3	8	0.21	0.8876
A	4	72	114124	<.0001
Site*A	12	72	0.93	0.5187
B	1	72	33205.5	<.0001
Site*B	3	72	0.18	0.9114
A*B	4	72	2585.98	<.0001
Site*A*B	12	72	0.92	0.5273
C	3	240	2122.50	<.0001

A*C	12	240	714.83	<.0001	
B*C	3	240	2620.89	<.0001	
A*B*C	12	240	1151.28	<.0001	
Site*C	9	240	1.74	0.0806	
Site*A*C	36	240	0.95	0.5530	
Site*B*C	9	240	2.23	0.0207	
Site*A*B*C	36	240	0.75	0.8450	

Least Squares Means

Effect	A	B	Site	Estimate	Standard Error	DF	t Value	Pr>|t|	Alpha
A	A1			7498.98	43.1792	72	173.67	<.0001	.
A	A2			8810.66	43.1792	72	204.05	<.0001	.
A	A3			10017	43.1792	72	232.00	<.0001	.
A	A4			11193	43.1792	72	259.22	<.0001	.
A	A5			12317	43.1792	72	285.25	<.0001	.
B		B1		9508.78	42.9586	72	221.35	<.0001	.
B		B2		10426	42.9586	72	242.70	<.0001	.
Site*A	A1		1	7460.92	86.3583	72	86.39	<.0001	.
Site*A	A2		1	8764.12	86.3583	72	101.49	<.0001	.
Site*A	A3		1	9965.12	86.3583	72	115.39	<.0001	.
Site*A	A4		1	11165	86.3583	72	129.29	<.0001	.
Site*A	A5		1	12279	86.3583	72	142.19	<.0001	.
Site*A	A1		2	7485.46	86.3583	72	86.68	<.0001	.
Site*A	A2		2	8785.83	86.3583	72	101.74	<.0001	.
Site*A	A3		2	9994.62	86.3583	72	115.73	<.0001	.
Site*A	A4		2	11158	86.3583	72	129.21	<.0001	.
Site*A	A5		2	12279	86.3583	72	142.19	<.0001	.
Site*A	A1		3	7528.33	86.3583	72	87.18	<.0001	.
Site*A	A2		3	8857.37	86.3583	72	102.57	<.0001	.
Site*A	A3		3	10062	86.3583	72	116.51	<.0001	.
Site*A	A4		3	11215	86.3583	72	129.86	<.0001	.
Site*A	A5		3	12354	86.3583	72	143.06	<.0001	.
Site*A	A1		4	7521.21	86.3583	72	87.09	<.0001	.
Site*A	A2		4	8835.29	86.3583	72	102.31	<.0001	.
Site*A	A3		4	10048	86.3583	72	116.35	<.0001	.
Site*A	A4		4	11234	86.3583	72	130.09	<.0001	.
Site*A	A5		4	12354	86.3583	72	143.06	<.0001	.
A	A1			7498.98	43.1792	72	173.67	<.0001	0.05
A	A2			8810.66	43.1792	72	204.05	<.0001	0.05
A	A3			10017	43.1792	72	232.00	<.0001	0.05
A	A4			11193	43.1792	72	259.22	<.0001	0.05
A	A5			12317	43.1792	72	285.25	<.0001	0.05
B		B1		9508.78	42.9586	72	221.35	<.0001	0.05
B		B2		10426	42.9586	72	242.70	<.0001	0.05
Site*A	A1		1	7460.92	86.3583	72	86.39	<.0001	0.05
Site*A	A2		1	8764.12	86.3583	72	101.49	<.0001	0.05
Site*A	A3		1	9965.12	86.3583	72	115.39	<.0001	0.05

APPENDIX

Effect	A							
Site*A	A4	1	11165	86.3583	72	129.29	<.0001	0.05
Site*A	A5	1	12279	86.3583	72	142.19	<.0001	0.05
Site*A	A1	2	7485.46	86.3583	72	86.68	<.0001	0.05
Site*A	A2	2	8785.83	86.3583	72	101.74	<.0001	0.05
Site*A	A3	2	9994.62	86.3583	72	115.73	<.0001	0.05
Site*A	A4	2	11158	86.3583	72	129.21	<.0001	0.05
Site*A	A5	2	12279	86.3583	72	142.19	<.0001	0.05
Site*A	A1	3	7528.33	86.3583	72	87.18	<.0001	0.05
Site*A	A2	3	8857.37	86.3583	72	102.57	<.0001	0.05
Site*A	A3	3	10062	86.3583	72	116.51	<.0001	0.05
Site*A	A4	3	11215	86.3583	72	129.86	<.0001	0.05
Site*A	A5	3	12354	86.3583	72	143.06	<.0001	0.05
Site*A	A1	4	7521.21	86.3583	72	87.09	<.0001	0.05
Site*A	A2	4	8835.29	86.3583	72	102.31	<.0001	0.05
Site*A	A3	4	10048	86.3583	72	116.35	<.0001	0.05
Site*A	A4	4	11234	86.3583	72	130.09	<.0001	0.05
Site*A	A5	4	12354	86.3583	72	143.06	<.0001	0.05

Least Squares Means

Effect	A	B	Site	Lower	Upper
A	A1			7412.90	7585.06
A	A2			8724.58	8896.73
A	A3			9931.31	10103
A	A4			11107	11279
A	A5			12231	12403
B		B1		9423.14	9594.42
B		B2		10340	10512
Site*A	A1		1	7288.76	7633.07
Site*A	A2		1	8591.97	8936.28
Site*A	A3		1	9792.97	10137
Site*A	A4		1	10993	11337
Site*A	A5		1	12107	12451
Site*A	A1		2	7313.31	7657.61
Site*A	A2		2	8613.68	8957.99
Site*A	A3		2	9822.47	10167
Site*A	A4		2	10986	11331
Site*A	A5		2	12107	12451
Site*A	A1		3	7356.18	7700.49
Site*A	A2		3	8685.22	9029.53
Site*A	A3		3	9889.43	10234
Site*A	A4		3	11042	11387
Site*A	A5		3	12182	12526
Site*A	A1		4	7349.06	7693.36
Site*A	A2		4	8663.14	9007.44
Site*A	A3		4	9876.06	10220
Site*A	A4		4	11062	11406
Site*A	A5		4	12182	12526

Differences of Least Squares Means

Effect	A	B	Site	_A	_B	_Site	Estimate	Standard Error	DF	t Value
A	A1			A2			−1311.68	7.9580	72	−164.82
A	A1			A3			−2518.41	7.9580	72	−316.46
A	A1			A4			−3694.10	7.9580	72	−464.20
A	A1			A5			−4817.69	7.9580	72	−605.39

Differences of Least Squares Means

Effect	A	B	Site	_A	_B	_Site	Pr>|t|	Adjustment	Adj P	Alpha
A	A1			A2			<.0001	Tukey-Kramer	<.0001	0.05
A	A1			A3			<.0001	Tukey-Kramer	<.0001	0.05
A	A1			A4			<.0001	Tukey-Kramer	<.0001	0.05
A	A1			A5			<.0001	Tukey-Kramer	<.0001	0.05

Differences of Least Squares Means

Effect	A	B	Site	_A	_B	_Site	Lower	Upper	Adj Lower	Adj Upper
A	A1			A2			−1327.54	−1295.81	.	.
A	A1			A3			−2534.27	−2502.54	.	.
A	A1			A4			−3709.97	−3678.24	.	.
A	A1			A5			−4833.55	−4801.82	.	.

Differences of Least Squares Means

Effect	A	B	Site	_A	_B	_Site	Estimate	Standard Error	DF	t Value
A	A2			A3			−1206.73	7.9580	72	−151.64
A	A2			A4			−2382.43	7.9580	72	−299.37
A	A2			A5			−3506.01	7.9580	72	−440.56
A	A3			A4			−1175.70	7.9580	72	−147.74
A	A3			A5			−2299.28	7.9580	72	−288.93
A	A4			A5			−1123.58	7.9580	72	−141.19
B		B1			B2		−917.15	5.0331	72	−182.22
Site*A	A1		1	A2		1	−1303.21	15.9161	72	−81.88
Site*A	A1		1	A3		1	−2504.21	15.9161	72	−157.34
Site*A	A1		1	A4		1	−3704.33	15.9161	72	−232.74
Site*A	A1		1	A5		1	−4818.25	15.9161	72	−302.73
Site*A	A1		1	A1		2	−24.5417	122.13	72	−0.20

.
.
.

The Mixed Procedure
Differences of Least Squares Means

Effect	A	B	Site	_A	_B	_Site	Pr>|t|	Adjustment	Adj P	Alpha
A	A2			A3			<.0001	Tukey-Kramer	<.0001	0.05
A	A2			A4			<.0001	Tukey-Kramer	<.0001	0.05
A	A2			A5			<.0001	Tukey-Kramer	<.0001	0.05
A	A3			A4			<.0001	Tukey-Kramer	<.0001	0.05

APPENDIX

A	A3		A5		<.0001	Tukey-Kramer <.0001	0.05
A	A4		A5		<.0001	Tukey-Kramer <.0001	0.05
B		B1		B2	<.0001	Tukey-Kramer <.0001	0.05
Site*A	A1	1	A2	1	<.0001	Tukey-Kramer <.0001	0.05
Site*A	A1	1	A3	1	<.0001	Tukey-Kramer <.0001	0.05
Site*A	A1	1	A4	1	<.0001	Tukey-Kramer <.0001	0.05
Site*A	A1	1	A5	1	<.0001	Tukey-Kramer <.0001	0.05
Site*A	A1	1	A1	2	0.8413	Tukey-Kramer 1.0000	0.05
Site*A	A1	1	A2	2	<.0001	Tukey-Kramer <.0001	0.05
.							
.							
.							
Site*A	A2	4	A4	4	<.0001	Tukey-Kramer <.0001	0.05
Site*A	A2	4	A5	4	<.0001	Tukey-Kramer <.0001	0.05
Site*A	A3	4	A4	4	<.0001	Tukey-Kramer <.0001	0.05
Site*A	A3	4	A5	4	<.0001	Tukey-Kramer <.0001	0.05
Site*A	A4	4	A5	4	<.0001	Tukey-Kramer <.0001	0.05

Differences of Least Squares Means

Effect	A	B	Site	_A	_B	_Site	Lower	Upper	Adj Lower	Adj Upper
Site*A	A2		4	A4		4	−2430.48	−2367.02	.	.
Site*A	A2		4	A5		4	−3550.60	−3487.15	.	.
Site*A	A3		4	A4		4	−1217.56	−1154.11	.	.
Site*A	A3		4	A5		4	−2337.69	−2274.23	.	.
Site*A	A4		4	A5		4	−1151.85	−1088.40	.	.

APPENDIX 10.2. EXAMPLE 10.2.

Factors A (5 levels) and B (8 levels) are in a split block arrangement in four blocks (replicates) at three sites. The complete data set is included on the book's FTP site and a SAS program for the data is presented below:

```
/*
Simulated data: split block across sites
whole plot factor: A (5 levels) is in a randomized complete block
design
split block factor: B (8 levels) is in a randomized complete
block design across levels of A per block.
number of sites = 3
number of blocks = 4
Numerical example: Section 10.3
*/

options ls = 79 nocenter nodate pageno = 1;
data spbsite;
input Obs Site A $ B $ Block $ Yield;
```

```
datalines;

  1  1  A1  B1  R1  8038
  2  1  A2  B1  R1  8231
  3  1  A3  B1  R1  8475
  4  1  A4  B1  R1  8606
  5  1  A5  B1  R1  8774
  6  1  A1  B2  R1  9031
     .
     .
     .
476  3  A1  B8  R4 13382
477  3  A2  B8  R4 13542
478  3  A3  B8  R4 13748
479  3  A4  B8  R4 13834
480  3  A5  B8  R4 13980


proc glm data = spbsite;
class Site A B Block;
model Yield = Site Block(Site)A A*site A*Block(Site) B B*site
B*Block(Site) A*B A*B*site;
random Site Block(Site) A*Block(Site) B*Block(Site)/test;
lsmeans A;
lsmeans B;
lmeans A*B A*site;


* multiple comparisons of means using tukey;
means A/tukey alpha = 0.05 e = A*Block(Site);
means B/tukey alpha = 0.05 e = B*Block(Site);
run;


* Using proc mixed;
proc mixed data = spbsite;
class Site A B Block;
model Yield = site A A*site B B*site A*B A*B*site;
random Block(Site) A*Block(Site) B*Block(site);

* LS means per factor;
lsmeans A;
lsmeans B;


* multiple comparisons of means using tukey;
lsmeans A/adjust = tukey alpha = 0.05;
lsmeans B/adjust = tukey alpha = 0.05;
lsmeans A*B/adjust = tukey alpha = 0.05;
run;
```

APPENDIX

The output for the above code and data in a drastically abbreviated form is presented below. Type I and III sums of squares are identical and the Type III ANOVA is omitted.

The GLM Procedure

```
         Class Level Information
Class     Levels     Values
Site         3       1 2 3
A            5       A1 A2 A3 A4 A5
B            8       B1 B2 B3 B4 B5 B6 B7 B8
Block        4       R1 R2 R3 R4
Number of observations 480
```

Dependent Variable: Yield

Source	DF	Sum of Squares	Mean Square	F Value	Pr > F
Model	227	6370995084	28066058	10813.6	<.0001
Error	252	654049	2595		
Corrected Total	479	6371649132			

R-Square	Coeff Var	Root MSE	Yield Mean
0.999897	0.459669	50.94537	11083.06

Source	DF	Type I SS	Mean Square	F Value	Pr > F
Site	2	523573968	261786984	100865	<.0001
Block(Site)	9	3756646710	417405190	160823	<.0001
A	4	29288163	7322041	2821.13	<.0001
Site*A	8	247899	30987	11.94	<.0001
A*Block(Site)	36	1783391	49539	19.09	<.0001
B	7	1937592291	276798899	106649	<.0001
Site*B	14	15903698	1135978	437.68	<.0001
B*Block(Site)	63	105727288	1678211	646.60	<.0001
A*B	28	91141	3255	1.25	0.1838
Site*A*B	56	140534	2510	0.97	0.5461

Source	Type III Expected Mean Square
Site	Var(Error) + 5 Var(B*Block(Site)) + 8 Var(A*Block(Site)) + 40 Var(Block(Site)) + 160 Var(Site) + Q(Site*A,Site*B,Site*A*B)
Block (Site)	Var(Error) + 5 Var(B*Block(Site)) + 8 Var(A*Block(Site)) + 40 Var(Block(Site))
A	Var(Error) + 8 Var(A*Block(Site)) + Q(A,Site*A,A*B,Site*A*B)
Site*A	Var(Error) + 8 Var(A*Block(Site)) + Q(Site*A,Site*A*B)
A*Block (Site)	Var(Error) + 8 Var(A*Block(Site))
B	Var(Error) + 5 Var(B*Block(Site)) + Q(B,Site*B,A*B,Site*A*B)

Site*B   Var(Error) + 5 Var(B*Block(Site)) + Q(Site*B, Site*A*B)
B*Block   Var(Error) + 5 Var(B*Block(Site))
 (Site)
A*B   Var(Error) + Q(A*B, Site*A*B)
Site*A*B   Var(Error) + Q(Site*A*B)

Tests of Hypotheses for Mixed Model Analysis of Variance

Dependent Variable: Yield

Source	DF	Type III SS	Mean Square	F Value	Pr>F
* Site	2	523573968	261786984	0.63	0.5559
Error	9	3756646710	417405190		

Error: MS(Block(Site))
* This test assumes one or more other fixed effects are zero.

Source	DF	Type III SS	Mean Square	F Value	Pr>F
Block(Site)	9	3756646710	417405190	241.95	<.0001
Error	66.472	114675113	1725154		

Error: MS(A*Block(Site)) + MS(B*Block(Site)) − MS(Error)

Source	DF	Type III SS	Mean Square	F Value	Pr>F
* A	4	29288163	7322041	147.80	<.0001
* Site*A	8	247899	30987	0.63	0.7508
Error	36	1783391	49539		

Error: MS(A*Block(Site))
* This test assumes one or more other fixed effects are zero.

Source	DF	Type III SS	Mean Square	F Value	Pr > F
A*Block(Site)	36	1783391	49539	19.09	<.0001
B*Block(Site)	63	105727288	1678211	646.60	<.0001
* A*B	28	91141	3255.042113	1.25	0.1838
Site*A*B	56	140534	2509.541667	0.97	0.5461
Error: MS(Error)	252	654049	2595.430704		

* This test assumes one or more other fixed effects are zero.

Source	DF	Type III SS	Mean Square	F Value	Pr > F
* B	7	1937592291	276798899	164.94	<.0001
* Site*B	14	15903698	1135978	0.68	0.7878
Error	63	105727288	1678211		

Error: MS(B*Block(Site))
* This test assumes one or more other fixed effects are zero.

Least Squares Means

A	Yield LSMEAN
A1	10740.1979
A2	10901.1250
A3	11081.9167
A4	11256.2708
A5	11435.8021

APPENDIX

```
Least Squares Means
B       Yield LSMEAN
B1        8016.6333
B2        8887.5833
B3        9773.5000
B4       10640.3500
B5       11520.6333
B6       12391.4500
B7       13282.0333
B8       14152.3167

Least Squares Means
A    B     Yield LSMEAN
A1   B1      7672.7500
A1   B2      8539.2500
A1   B3      9429.5833
A1   B4     10288.2500
A1   B5     11177.8333
A1   B6     12051.3333
A1   B7     12939.5000
A1   B8     13823.0833
A2   B1      7833.9167
 .
 .
 .
A5   B6     12754.3333
A5   B7     13630.0000
A5   B8     14479.3333

Site  A     Yield LSMEAN
 1    A1     10369.9688
 1    A2     10575.9688
 1    A3     10755.5625
 1    A4     10941.1875
 1    A5     11129.6563
 2    A1      9707.5313
 2    A2      9826.9375
 2    A3     10002.4688
 2    A4     10151.5000
 2    A5     10313.0000
 3    A1     12143.0938
 3    A2     12300.4688
 3    A3     12487.7188
 3    A4     12676.1250
 3    A5     12864.7500

Tukey's Studentized Range (HSD) Test for Yield
Alpha                              0.05
Error Degrees of Freedom             36
```

Error Mean Square   49538.65
Critical Value of Studentized Range   4.05997
Minimum Significant Difference   92.227

Means with the same letter are not significantly different.

T
u
k
.
.
.
g   Mean   N   A
A   11435.80 96   A5
B   11256.27 96   A4
C   11081.92 96   A3
D   10901.13 96   A2
E   10740.20 96   A1

Tukey's Studentized Range (HSD) Test for Yield

Alpha   0.05
Error Degrees of Freedom   63
Error Mean Square   1678211
Critical Value of Studentized Range   4.43360
Minimum Significant Difference   741.49

Means with the same letter are not significantly different.

     Mean   N   B
A   14152.3 60   B8
B   13282.0 60   B7
C   12391.5 60   B6
D   11520.6 60   B5
E   10640.4 60   B4
F   9773.5 60   B3
G   8887.6 60   B2
H   8016.6 60   B1

The Mixed Procedure
  Covariance Parameter
      Estimates
Cov Parm   Estimate
Site   0
Block(Site)   10391884
A*Block(Site)   5867.83
B*Block(Site)   335119
Residual   2595.44

APPENDIX

Type 3 Tests of Fixed Effects

Effect	Num DF	Den DF	F Value	Pr>F
A	4	36	147.81	<.0001
Site*A	8	36	0.63	0.7508
B	7	63	164.94	<.0001
Site*B	14	63	0.68	0.7878
A*B	28	252	1.25	0.1838
Site*A*B	56	252	0.97	0.5461

Least Squares Means

Effect	A	B	Estimate	Standard Error	DF	t Value	Pr>|t|	Alpha
A	A1		10740	932.74	36	11.51	<.0001	0.05
A	A2		10901	932.74	36	11.69	<.0001	0.05
A	A3		11082	932.74	36	11.88	<.0001	0.05
A	A4		11256	932.74	36	12.07	<.0001	0.05
A	A5		11436	932.74	36	12.26	<.0001	0.05
B		B1	8016.63	945.55	63	8.48	<.0001	0.05
B		B2	8887.58	945.55	63	9.40	<.0001	0.05
B		B3	9773.50	945.55	63	10.34	<.0001	0.05
B		B4	10640	945.55	63	11.25	<.0001	0.05
B		B5	11521	945.55	63	12.18	<.0001	0.05
B		B6	12391	945.55	63	13.11	<.0001	0.05
B		B7	13282	945.55	63	14.05	<.0001	0.05
B		B8	14152	945.55	63	14.97	<.0001	0.05
A*B	A1	B1	7672.75	945.84	252	8.11	<.0001	0.05
A*B	A1	B2	8539.25	945.84	252	9.03	<.0001	0.05
A*B	A1	B3	9429.58	945.84	252	9.97	<.0001	0.05
A*B	A1	B4	10288	945.84	252	10.88	<.0001	0.05
A*B	A1	B5	11178	945.84	252	11.82	<.0001	0.05
A*B	A1	B6	12051	945.84	252	12.74	<.0001	0.05
A*B	A1	B7	12939	945.84	252	13.68	<.0001	0.05
A*B	A1	B8	13823	945.84	252	14.61	<.0001	0.05
A*B	A2	B1	7833.92	945.84	252	8.28	<.0001	0.05
A*B	A2	B2	8706.42	945.84	252	9.20	<.0001	0.05
A*B	A2	B3	9574.50	945.84	252	10.12	<.0001	0.05
A*B	A2	B4	10488	945.84	252	11.09	<.0001	0.05
A*B	A2	B5	11321	945.84	252	11.97	<.0001	0.05
A*B	A2	B6	12197	945.84	252	12.90	<.0001	0.05
A*B	A2	B7	13100	945.84	252	13.85	<.0001	0.05
A*B	A2	B8	13988	945.84	252	14.79	<.0001	0.05
A*B	A3	B1	8032.67	945.84	252	8.49	<.0001	0.05
A*B	A3	B2	8877.67	945.84	252	9.39	<.0001	0.05
A*B	A3	B3	9761.00	945.84	252	10.32	<.0001	0.05
A*B	A3	B4	10605	945.84	252	11.21	<.0001	0.05

Effect	A	B		Estimate	Std Error	DF	t Value	Pr>|t|	
A*B	A3	B5		11548	945.84	252	12.21	<.0001	0.05
A*B	A3	B6		12394	945.84	252	13.10	<.0001	0.05
A*B	A3	B7		13284	945.84	252	14.04	<.0001	0.05

Least Squares Means

Effect	A	B	Lower	Upper
A	A1		8848.52	12632
A	A2		9009.45	12793
A	A3		9190.24	12974
A	A4		9364.59	13148
A	A5		9544.13	13327
B		B1	6127.11	9906.16
B		B2	6998.06	10777
B		B3	7883.98	11663
B		B4	8750.83	12530
B		B5	9631.11	13410
B		B6	10502	14281
B		B7	11393	15172
B		B8	12263	16042
A*B	A1	B1	5809.98	9535.52
A*B	A1	B2	6676.48	10402
A*B	A1	B3	7566.82	11292
A*B	A1	B4	8425.48	12151
A*B	A1	B5	9315.07	13041
A*B	A1	B6	10189	13914
A*B	A1	B7	11077	14802
A*B	A1	B8	11960	15686
.				
.				
.				
A*B	A5	B7	11767	15493
A*B	A5	B8	12617	16342

Differences of Least Squares Means

Effect	A	B	_A	_B	Estimate	Standard Error	DF	t Value	Pr>|t|
A	A1		A2		−160.93	32.1254	36	−5.01	<.0001
A	A1		A3		−341.72	32.1254	36	−10.64	<.0001
A	A1		A4		−516.07	32.1254	36	−16.06	<.0001
A	A1		A5		−695.60	32.1254	36	−21.65	<.0001
A	A2		A3		−180.79	32.1254	36	−5.63	<.0001
A	A2		A4		−355.15	32.1254	36	−11.05	<.0001
A	A2		A5		−534.68	32.1254	36	−16.64	<.0001
A	A3		A4		−174.35	32.1254	36	−5.43	<.0001
A	A3		A5		−353.89	32.1254	36	−11.02	<.0001
A	A4		A5		−179.53	32.1254	36	−5.59	<.0001
B		B1		B2	−870.95	236.52	63	−3.68	0.0005
B		B1		B3	−1756.87	236.52	63	−7.43	<.0001
B		B1		B4	−2623.72	236.52	63	−11.09	<.0001

APPENDIX

B		B1		B5	−3504.00	236.52	63	−14.82	<.0001
B		B1		B6	−4374.82	236.52	63	−18.50	<.0001
B		B1		B7	−5265.40	236.52	63	−22.26	<.0001
B		B1		B8	−6135.68	236.52	63	−25.94	<.0001
B		B2		B3	−885.92	236.52	63	−3.75	0.0004
B		B2		B4	−1752.77	236.52	63	−7.41	<.0001
B		B2		B5	−2633.05	236.52	63	−11.13	<.0001
B		B2		B6	−3503.87	236.52	63	−14.81	<.0001
B		B2		B7	−4394.45	236.52	63	−18.58	<.0001
B		B2		B8	−5264.73	236.52	63	−22.26	<.0001
B		B3		B4	−866.85	236.52	63	−3.67	0.0005
B		B3		B5	−1747.13	236.52	63	−7.39	<.0001
B		B3		B6	−2617.95	236.52	63	−11.07	<.0001
B		B3		B7	−3508.53	236.52	63	−14.83	<.0001
B		B3		B8	−4378.82	236.52	63	−18.51	<.0001
B		B4		B5	−880.28	236.52	63	−3.72	0.0004
B		B4		B6	−1751.10	236.52	63	−7.40	<.0001
B		B4		B7	−2641.68	236.52	63	−11.17	<.0001
B		B4		B8	−3511.97	236.52	63	−14.85	<.0001
B		B5		B6	−870.82	236.52	63	−3.68	0.0005
B		B5		B7	−1761.40	236.52	63	−7.45	<.0001
B		B5		B8	−2631.68	236.52	63	−11.13	<.0001
B		B6		B7	−890.58	236.52	63	−3.77	0.0004
B		B6		B8	−1760.87	236.52	63	−7.45	<.0001
B		B7		B8	−870.28	236.52	63	−3.68	0.0005
A*B	A1	B1	A1	B2	−866.50	237.25	252	−3.65	0.0003
A*B	A1	B1	A1	B3	−1756.83	237.25	252	−7.41	<.0001
A*B	A1	B1	A1	B4	−2615.50	237.25	252	−11.02	<.0001
A*B	A1	B1	A1	B5	−3505.08	237.25	252	−14.77	<.0001
A*B	A1	B1	A1	B6	−4378.58	237.25	252	−18.46	<.0001
A*B	A1	B1	A1	B7	−5266.75	237.25	252	−22.20	<.0001
A*B	A1	B1	A1	B8	−6150.33	237.25	252	−25.92	<.0001
A*B	A1	B1	A2	B1	−161.17	37.5572	252	−4.29	<.0001
A	A1		A2		Tukey-Kramer	0.0001	0.05	−226.08	−95.7737
A	A1		A3		Tukey-Kramer	<.0001	0.05	−406.87	−276.57
A	A1		A4		Tukey-Kramer	<.0001	0.05	−581.23	−450.92
A	A1		A5		Tukey-Kramer	<.0001	0.05	−760.76	−630.45
A	A2		A3		Tukey-Kramer	<.0001	0.05	−245.95	−115.64
A	A2		A4		Tukey-Kramer	<.0001	0.05	−420.30	−289.99
A	A2		A5		Tukey-Kramer	<.0001	0.05	−599.83	−469.52
A	A3		A4		Tukey-Kramer	<.0001	0.05	−239.51	−109.20
A	A3		A5		Tukey-Kramer	<.0001	0.05	−419.04	−288.73
A	A4		A5		Tukey-Kramer	<.0001	0.05	−244.68	−114.38
B		B1		B2	Tukey-Kramer	0.0107	0.05	−1343.59	−398.31
B		B1		B3	Tukey-Kramer	<.0001	0.05	−2229.51	−1284.23
B		B1		B4	Tukey-Kramer	<.0001	0.05	−3096.36	−2151.08
B		B1		B5	Tukey-Kramer	<.0001	0.05	−3976.64	−3031.36
B		B1		B6	Tukey-Kramer	<.0001	0.05	−4847.46	−3902.18
B		B1		B7	Tukey-Kramer	<.0001	0.05	−5738.04	−4792.76

B		B1		B8 Tukey-Kramer	<.0001	0.05	-6608.32 -5663.04
B		B2		B3 Tukey-Kramer	0.0088	0.05	-1358.56 -413.28
B		B2		B4 Tukey-Kramer	<.0001	0.05	-2225.41 -1280.13
B		B2		B5 Tukey-Kramer	<.0001	0.05	-3105.69 -2160.41
B		B2		B6 Tukey-Kramer	<.0001	0.05	-3976.51 -3031.23
B		B2		B7 Tukey-Kramer	<.0001	0.05	-4867.09 -3921.81
B		B2		B8 Tukey-Kramer	<.0001	0.05	-5737.37 -4792.09
B		B3		B4 Tukey-Kramer	0.0113	0.05	-1339.49 -394.21
B		B3		B5 Tukey-Kramer	<.0001	0.05	-2219.77 -1274.49
B		B3		B6 Tukey-Kramer	<.0001	0.05	-3090.59 -2145.31
B		B3		B7 Tukey-Kramer	<.0001	0.05	-3981.17 -3035.89
B		B3		B8 Tukey-Kramer	<.0001	0.05	-4851.46 -3906.18
B		B4		B5 Tukey-Kramer	0.0095	0.05	-1352.92 -407.64
B		B4		B6 Tukey-Kramer	<.0001	0.05	-2223.74 -1278.46
B		B4		B7 Tukey-Kramer	<.0001	0.05	-3114.32 -2169.04
B		B4		B8 Tukey-Kramer	<.0001	0.05	-3984.61 -3039.33
B		B5		B6 Tukey-Kramer	0.0107	0.05	-1343.46 -398.18
B		B5		B7 Tukey-Kramer	<.0001	0.05	-2234.04 -1288.76
B		B5		B8 Tukey-Kramer	<.0001	0.05	-3104.32 -2159.04
B		B6		B7 Tukey-Kramer	0.0083	0.05	-1363.22 -417.94
B		B6		B8 Tukey-Kramer	<.0001	0.05	-2233.51 -1288.23
B		B7		B8 Tukey-Kramer	0.0108	0.05	-1342.92 -397.64
A*B	A1	B1	A1	B2 Tukey-Kramer	0.1204	0.05	-1333.74 -399.26
A*B	A1	B1	A1	B3 Tukey-Kramer	<.0001	0.05	-2224.07 -1289.60
A*B	A1	B1	A1	B4 Tukey-Kramer	<.0001	0.05	-3082.74 -2148.26
A*B	A1	B1	A1	B5 Tukey-Kramer	<.0001	0.05	-3972.32 -3037.85
A*B	A1	B1	A1	B6 Tukey-Kramer	<.0001	0.05	-4845.82 -3911.35
A*B	A1	B1	A1	B7 Tukey-Kramer	<.0001	0.05	-5733.99 -4799.51
A*B	A1	B1	A1	B8 Tukey-Kramer	<.0001	0.05	-6617.57 -5683.10
A*B	A1	B1	A2	B1 Tukey-Kramer	0.0142	0.05	-235.13 -87.2006

Most of the $40(39)/2 = 780$ pairwise comparisons have been deleted as has much of the additional output.

A*B	A5	B5	A5	B8 Tukey-Kramer	<.0001	0.05	-3077.49 -2143.01
A*B	A5	B6	A5	B7 Tukey-Kramer	0.1077	0.05	-1342.90 -408.43
A*B	A5	B6	A5	B8 Tukey-Kramer	<.0001	0.05	-2192.24 -1257.76
A*B	A5	B7	A5	B8 Tukey-Kramer	0.1475	0.05	-1316.57 -382.10

CHAPTER 11

Covariance Analyses for Split Plot and Split Block Experiment Designs

11.1. INTRODUCTION

In our discussion of covariance analyses for split plot and split block experiment designs, the presentation given by Federer and Meredith (1989, 1992) is followed herein. The concepts, population structure, philosophical nature, analyses, and usage of multiple error terms for the general families of split plot, split block, and variations of these as explained in previous chapters are not well explained or documented in statistical literature and statistical software (Federer, 1977). The standard split plot experiment design as discussed by Yates (1937) is described in Chapter 1. The subject of covariance for the experiment designs discussed in this text is quite limited in statistical literature. Exceptions are the discussions by Kempthorne (1952), Chapter 16 of Federer (1955), and Federer and Meredith (1989, 1992). The information available here was not used for a number of software packages as is indicated in Federer and Henderson (1978, 1979); Federer et al. (1987); Federer et al. (1987a, 1987b); Meredith et al. (1988); and Miles-McDermott et al. (1988). An exception is the GENSTAT software package.

There are several ways in which a covariate may be used as a treatment variable. For example, use a polynomial regression model to estimate linear, quadratic, and so forth trends for the various quantitative levels of a factor. A multivariate analysis could be performed such as presented by Steele and Federer (1955). Estimates of slope and curvature regressions could be obtained for each treatment and each treatment combination in an experiment. Milliken and Johnson (2002) treat a covariate as a treatment predictor variable. In Section 15.3, they consider a covariate measured on the whole plot and obtain linear regression coefficients for each combination of whole plot and split treatments. In Section 15.4, they do something

---

Variations on Split Plot and Split Block Experiment Designs, by Walter T. Federer and Freedom King
Copyright © 2007 John Wiley & Sons, Inc.

similar but with the covariate measured on the split plot. In Section 15.5, they present similar considerations for the case where one covariate is measured on the whole plot and a second covariate is measured on the split plot. These are investigations of the interaction of the treatments with the covariate. Simple linear (or higher order) regression models for one covariate are obtained for treatment variables only. They rule out consideration of multiple regressions with the blocking variables in their considerations as stated in Section 15.2.

The references in the first paragraph above use a covariate as a blocking variable and this is the approach presented in this chapter. Steele and Federer (1955) and Milliken and Johnson (2002) use a covariate as a treatment predictor variable. Either approach or a combination of the two approaches may be desirable in an analysis of data from experiments. There are several ways to use a covariate with blocking variables. For example, trends and curvatures in rows, in columns, in blocks, and interactions of trends have been used in trend analysis of data (See Federer, 2003). Spatial location or order within the experiment has been used as a covariate in trend analysis.

In an analysis of covariance, ANCOVA, *there are as many blocking variable regression coefficients as there are error terms in an analysis of variance*, ANOVA. Multiple error terms and multiple regression coefficients add complexity and difficulty to the statistical analysis and computer programming. Formulae for obtaining treatment means adjusted for a covariate and the variances of a difference between two such means are more complex and numerous. For the more complex experiment designs discussed in this text, it may be advisable to avoid using covariates as formulae for the means as their variances are mostly unavailable. If necessary to use a covariate, an alternative procedure would be to perform a covariance analysis for different subsets of the data that fit into one of the settings described in this chapter.

11.2. COVARIANCE ANALYSIS FOR A STANDARD SPLIT PLOT DESIGN

As described in Chapter 1, the almost universal split plot experiment design discussed in statistical textbooks consists of the whole plot treatments arranged in a randomized complete block experiment design and the split plot treatments randomly assigned to each of the whole plots. This is the standard split plot experiment design (Federer, 1955, 1975, 1977; Yates, 1937). As demonstrated herein, there exist many variations of split plot experiment designs used in practice. There are many different design structures for the whole plot treatments and for the split plot treatments.

ANCOVA will be described for the standard split plot experiment design. Depending upon the particular experimental situation, there are many models available for analyzing the responses from split plot designed experiments. We shall use the model described by Yates (1937) and Federer (1955). Let the *hij*th observation be denoted by the Y_{hij} observation and the Z_{hij} covariate. A response model equation with a covariate may be of the form (Kempthorne, 1952; Federer, 1955; Federer and Meredith, 1992):

$$Y_{hij} = \mu + \rho_h + \alpha_r + \delta_{hi} + \beta_w(\bar{z}_{hi.} - \bar{z}_{...}) + \gamma_j + \alpha\gamma_{ij} + \beta_s(Z_{hij} - \bar{z}_{hi.}) + \varepsilon_{hij}, \quad (11.1)$$

where

μ is a general mean effect,

ρ_h is the hth random block effect and is IID(0, σ_ρ^2),

α_i is the ith whole plot treatment effect, factor A,

γ_j is the jth split plot treatment effect, factor B,

δ_{hi} is the hith random whole plot error effect, and is IID(0, σ_δ^2),

$\alpha\gamma_{ij}$ is the ijth interaction effect of factors A and B,

ε_{hij} is the hijth random split plot error effect, and is IID(0, σ_ε^2),

$h = 1, 2, \ldots, r, i = 1, 2, \ldots, a$, and $j = 1, 2, \ldots, b$, and

$\bar{z}_{hi.}$ is the mean of Z_{hij} for the hith whole plot.

The dot summation means summed over the subscript replaced by the dot. The random effects in the model, ρ_h, δ_{hi}, and ε_{hij}, are assumed to be mutually independent.

The covariate value Z_{hij} in equation (11.1) is for the hijth split plot experimental unit, β_s is the split plot regression coefficient, and β_w is the whole plot regression coefficient. In some cases, covariates are for whole plot experimental units in which case the covariate in (11.1) is written as Z_{hi} The product of the estimated residual effects e_{hij}, for split plot experimental units and the Z_{hij} values of the covariate are used to obtain the split plot linear regression coefficient β_s. The product of the residual effects for whole plots d_{hi}, and the sum of the covariate values summed over j $Z_{hi.}$ form the linear whole plot regression coefficient β_w. There are two regressions in a split plot ANCOVA needed for adjusting the means of the response Y for values of the covariate Z_{hij}. As a word of caution, some computer software packages use the split plot regression coefficient for adjusting all the means (see above references.). In general, there will be as many regression terms as there are error terms in the ANCOVA. For example, if there are six error terms in an ANCOVA, there will be six regression coefficients. Some means may require the use of all six in order to adjust for a covariate while others would use only one regression coefficient. Owing to the algebraic complexity involved, analysts may forego the use of covariates for some mean combinations in the case of many error terms.

An ANCOVA table of the sums of squares and cross-products is presented in Table 11.1. If F-statistics are desired, they may be computed using the adjusted mean squares as follows:

$$\frac{A'_{yy}/(a-1)}{D'_{yy}/(ar-a-r)} = \frac{A'_{yy}/(a-1)}{E_a}, \quad (11.2)$$

Table 11.1. ANCOVA for a Standard Split Plot Experiment Design with a Covariate.

Source of variation	Degrees of freedom (DF)	Sums of products yy	yz	zz	DF	Adjusted sums of squares
Total	rab	T_{yy}	T_{yz}	T_{zz}		
Mean	1	M_{yy}	M_{yz}	M_{zz}		
Block, R	$r-1$	R_{yy}	R_{yz}	R_{zz}		
Factor A	$a-1$	A_{yy}	A_{yz}	A_{zz}		
Error A	$(a-1)(r-1)$	D_{yy}	D_{yz}	D_{zz}	$ra-r-a$	$D_{yy} - \dfrac{D_{yz}^2}{D_{zz}} = D'_{yy}$
Factor B	$b-1$	B_{yy}	B_{yz}	B_{zz}		
$A \times B$	$(a-1)(b-1)$	I_{yy}	I_{yz}	I_{zz}	$ab-a-b$	$I_{yy} - \dfrac{I_{yz}^2}{I_{zz}} = I'_{yy}$
Error B	$a(b-1)(r-1)$	E_{yy}	E_{yz}	E_{zz}	$a(b-1)(r-1)-1$	$E_{yy} - \dfrac{E_{yz}^2}{E_{zz}} = E'_{yy}$
A (adj. for β_w)	$a-1$		$A_{yy} - \dfrac{(A_{yz}+D_{yz})^2}{A_{zz}+D_{zz}} + \dfrac{D_{yz}^2}{D_{zz}} = A'_{yy}$			
B (adj. for β_s)	$b-1$		$B_{yy} - \dfrac{(B_{yz}+E_{yz})^2}{B_{zz}+E_{zz}} + \dfrac{E_{yz}^2}{E_{zz}} = B'_{yy}$			
$A \times B$ (adjusted for β_s)	$(a-1)(b-1)$		$I_{yy} - \dfrac{(I_{yz}+E_{yz})^2}{I_{zz}+E_{zz}} + \dfrac{E_{yz}^2}{E_{zz}} = I'_{yy}$			

$$\frac{B'_{yy}/(b-1)}{E'_{yy}/\{a(b-1)(r-1)-1\}} = \frac{B'_{yy}/(b-1)}{E_b}, \qquad (11.3)$$

and

$$\frac{I'_{yy}/(a-1)(b-1)}{E'_{yy}/\{a(b-1)(r-1)-1\}} = \frac{I'_{yy}/(a-1)(b-1)}{E_b}. \qquad (11.4)$$

Other F-statistics may involve the use of the $A \times B$ interaction as an error mean square. In that case, I'_{yy} is used in place of E'_{yy}. This means that there are three instead of two regressions to be considered.

The various sums of cross-products are obtained as given below:

$$T_{yz} = \sum_{h=1}^{r} \sum_{i=1}^{a} \sum_{j=1}^{b} Y_{hij} Z_{hij} \qquad (11.5)$$

$$M_{yz} = \frac{Y_{...} Z_{...}}{rab} \qquad (11.6)$$

$$R_{yz} = \frac{\sum_{h=1}^{r} Y_{h..} Z_{h..}}{ab} - M_{yz} \qquad (11.7)$$

$$A_{yz} = \frac{\sum_{i=1}^{a} Y_{.i.} Z_{.i.}}{rb} - M_{yz} \qquad (11.8)$$

COVARIANCE ANALYSIS FOR A STANDARD SPLIT PLOT DESIGN

$$D_{yz} = \frac{\sum_{h=1}^{r} \sum_{i=1}^{a} Y_{hi.} Z_{hi.}}{b} - R_{yz} - A_{yz} + M_{yz} \qquad (11.9)$$

$$B_{yz} = \frac{\sum_{j=1}^{b} Y_{..j} Z_{..j}}{ar} - M_{yz} \qquad (11.10)$$

$$I_{yz} = \frac{\sum_{i=1}^{a} \sum_{j=1}^{b} Y_{.ij} Z_{.ij}}{r} - A_{yz} - B_{yz} + M_{yz}. \qquad (11.11)$$

The Error B sum of cross-products is obtained by subtraction. The computations in equations (11.5) to (11.11) with adjustments for unequal numbers hold for non-orthogonal or unbalanced data sets as well. The mean squares in an ANCOVA are obtained by dividing with the appropriate degrees of freedom. The ANCOVA case discussed here is for one covariate and a common slope, but the extension to more that one covariate or for regressions other than linear is straightforward.

The various Y means are adjusted (adj.) for the covariate Z as follows:

$$\bar{y}_{.ij}(\text{adj.}) = \bar{y}_{.ij} - \hat{\beta}_s(\bar{z}_{.ij} - \bar{z}_{.i.}) - \hat{\beta}_w(\bar{z}_{.i.} - \bar{z}_{...}) = \bar{y}'_{.ij}, \qquad (11.12)$$

$$\bar{y}_{..j}(\text{adj.}) = \bar{y}_{..j} - \hat{\beta}_s(\bar{z}_{..j} - \bar{z}_{...}) = \bar{y}'_{..j}, \qquad (11.13)$$

and

$$\bar{y}_{.i.}(\text{adj.}) = \bar{y}_{.i.} - \hat{\beta}_w(\bar{z}_{.i.} - \bar{z}_{...}) = \bar{y}'_{.i.}. \qquad (11.14)$$

Estimates of β_w and β_s, $\hat{\beta}_w = D_{yz}/D_{zz}$ and $\hat{\beta}_s = E_{yz}/E_{zz}$, respectively, are used in the above formulae when adjusting the means for the covariate.

Estimated variances of a difference between two estimated adjusted means for $i \neq i'$ and $j \neq j'$ are as follows:

Variance of a difference between two adjusted whole plot means:

$$V(\bar{y}'_{.i.} - \bar{y}'_{.i'.}) = E_a \left[\frac{2}{br} + \frac{(\bar{z}_{.i.} - \bar{z}_{.i'.})^2}{D_{zz}} \right]. \qquad (11.15)$$

Variance of a difference between two adjusted split plot means:

$$V(\bar{y}'_{..j} - \bar{y}'_{..j'}) = E_b \left[\frac{2}{ar} + \frac{(\bar{z}_{..j} - \bar{z}_{..j'})^2}{E_{zz}} \right]. \qquad (11.16)$$

Variance of a difference between two adjusted split plot means for the same whole plot:

$$V(\bar{y}'_{.ij} - \bar{y}'_{.ij'}) = E_b \left[\frac{2}{r} + \frac{(\bar{z}_{.ij} - \bar{z}_{.ij'})^2}{E_{zz}} \right]. \qquad (11.17)$$

Variance of a difference between two adjusted whole plot means for the same split plot:

$$V(\bar{y}'_{.ij} - \bar{y}'_{.i'j}) = \frac{2(E_b + \sigma_\delta^2)}{r} + \frac{E_a(\bar{z}_{.i.} - \bar{z}_{.i'.})^2}{D_{zz}} + \frac{E_b(\bar{z}_{.ij} - \bar{z}_{.i'j} - \bar{z}_{.i.} + \bar{z}_{.i'.})^2}{E_{zz}}. \tag{11.18}$$

The estimated variance of a difference between two different split plot means from two different whole plot treatments is the same as the last variance. In the above formulae,

$$E_a = \hat{\sigma}_\varepsilon^2 + b\hat{\sigma}_\delta^2 = \frac{D'_{yy}}{ar - r - a}, \tag{11.19}$$

$$\hat{\sigma}_\varepsilon^2 = \frac{E'_{yy}}{a(b-1)(r-1) - 1} = E_b, \tag{11.20}$$

and

$$\hat{\sigma}_\delta^2 = \frac{(E_a - E_b)}{b}. \tag{11.21}$$

The hat indicates an estimated quantity. E_a is associated with $ar - a - r$ degrees of freedom and E_b is associated with $a(b-1)(r-1) - 1$ degrees of freedom. The degrees of freedom, f^*, for the variance $E_b + \sigma_\delta^{2*} = [(b-1)E_b + E_a]/b$ needs to be approximated using the formula given below:

$$t_\alpha(f^*) = \frac{(E_a t_{\alpha, ar-a-r=df} + (b-1)E_b t_{\alpha, a(b-1)(r-1)-1=df})}{(b-1)E_b + E_a}, \tag{11.22}$$

where $t_\alpha(f^*)$ is the tabulated value of the t statistic at the α percentage level for f^* degrees of freedom. This approximation generally underestimates the degrees of freedom for variances constructed in this manner (See Cochran and Cox, 1957 and Grimes and Federer, 1984). Other methods for approximating the degrees of freedom for constructed variances and variance components are available in the literature.

Example 11.1: Covariate varies with split plots—Hypothetical data are used to illustrate the statistical analysis for ANCOVA for a split plot experiment design. The data set size is kept small in order to allow manual calculations for the analysis. Factor W, whole plot, has two levels, W1 and W2, and is in a randomized complete block arrangement in the three blocks. Factor S, split plot, has four levels, S1, S2, S3, and S4, and is randomized within each whole plot W. A SAS code for most of the analysis is given in Appendix 11.1. A comparison of the analysis herein and the output from the code will demonstrate what is available from the code output for ANCOVA. The data are given in Table 11.2. Various totals for the Y and Z variables

Table 11.2. Hypothetical Data Y from a Split Plot Experiment Design with a Covariate Z (Second Item in Each Pair).

Split plot	Whole plot 1 = W1					Whole plot 2 = W2				
	|S1	S2	S3	S4	|Total	|S1	S2	S3	S4	|Total
Block B1	|3 1	4 2	7 1	6 2	|20 6	|3 2	2 0	1 2	14 4	|20 8
B2	|6 2	10 2	1 0	11 4	|28 8	|8 4	8 1	2 3	18 4	|36 12
B3	|6 3	10 5	4 2	4 0	|24 10	|10 3	8 2	9 4	13 7	|40 16
Total	|15 6	24 9	12 3	21 6	|72 24	|21 9	18 3	12 9	45 15	|96 36

are given in Table 11.2. Other totals that are required in an analysis of covariance are given in Table 11.3. The sums of squares for variables Y and Z are obtained as described in Chapter 1. The detailed computations for the cross products of variables Y and Z are given by equations (11.5) to (11.11) and computed as indicated below. The format of the ANCOVA table, Table 11.4, follows that given for Table 11.1. Total sum of cross-products, equation (11.5) is as follows:

$$3(1) + 4(2) + 7(1) + 6(2) + \ldots + 10(3) + 8(2) + 9(4) + 13(7) = 537.$$

Correction for the mean cross-products, equation (11.6):

$$168(60)/24 = 420.$$

Block sum of cross-products, equation (11.7):

$$[40(14) + 64(20) + 64(26)]/[4(2) = 8] - 168(60)/24 = 18$$

W sum of cross-products, equation (11.8):

$$[72(24) + 96(36)]/[4(3) = 12] - 168(60)/24 = 12.$$

$R \times W$ sum of cross-products, equation (11.9):

$$[20(6) + 28(8) + 24(10) + 20(8) + 36(12) + 40(16)]/4 - 168(60)/24 - 18 - 12 = 4.$$

Table 11.3. Other Totals Used in ANCOVA.

		W1		W2		Total		Totals						
								S1		S2		S3		S4
Variable		Y	Z	Y	Z	Y	Z	Y	Z	Y	Z	Y	Z	Y	Z
Block	B1	20	6	20	8	40	14	36	15	42	12	24	12	66	21
	B2	28	8	36	12	64	20								
	B3	24	10	40	16	64	26								
Total		72	24	96	36	168	60								

Table 11.4. ANCOVA for the Data in Table 11.2.

Source of variation	Degrees of freedom (DF)	Sums of products			DF	Adj. sums of squares	Mean square
		yy	yz	zz			
Total	24	1,616	537	216			
Mean	1	1,176	420	150			
Block, R	2	48	18	9			
W	1	24	12	6			
W × R	2	16	4	1	1	0	0
S	3	156	33	9			
S × W	3	84	33	21			
S × R: W	12	112	17	20	11	97.55	8.87
W (adj. for β_w)					1	$W'_{yy} = 3.43$	3.43
S (adj. for β_s)					3	$S'_{yy} = 84.24$	28.08
S × W (adj. for β_s)					3	$I'_{yy} = 37.47$	12.49

S sum of cross-products, equation (11.10):

$$[36(15) + 42(12) + 24(12) + 66(21)]/[2(3) = 6] - 168(60)/24 = 33.$$

$S \times W$ interaction sum of cross-products, equations (11.11):

$$[15(6) + 24(9) + 12(3) + 21(6) + 21(9) + 18(3) + 12(9) + 45(15)]/3$$
$$- 168(60)/24 - 33 - 12 = 33.$$

$S \times R : W$ sum of cross-products:

$$\{3(1) + 6(2) + 6(3) + \ldots + 6(2) + 11(4) + 4(0) - [20(6) + 28(8) + 24(10)]/4$$
$$- [15(6) + 24(9) + 12(3) + 21(6)]/3 + 72(24)/12\} + 3(2) + 8(4) + 10(3)$$
$$+ \ldots + 14(4) + 18(4) + 13(7) - [20(8) + 36(12) + 40(16)]/4 - [21(9)$$
$$+ 18(3) + 12(9) + 45(15)]/3 + 96(36)/12 = 17.$$

$W + R \times W$ adjusted for whole plot regression with $1 + 2 - 1 = 2$ degrees of freedom:

$$24 + 16 - (12 + 4)^2/(6 + 1) = 40 - 256/7 = 3.43.$$

$R \times W$ adjusted for whole plot regression with $2 - 1 = 1$ degree of freedom:

$$16 - 4^2/1 = 0.$$

W adjusted for whole plot regression is the difference of the above two:

$$3.43 - 0 = 3.43.$$

COVARIANCE ANALYSIS FOR A STANDARD SPLIT PLOT DESIGN

$S + S \times R : W$ adjusted for split plot regression with $12 + 3 - 1 = 14$ degrees of freedom:

$$156 + 112 - (33 + 17)^2/(9 + 20) = 181.79.$$

$S \times W + S \times R : W$ adjusted for split plot regression with $12 + 3 - 1 = 14$ degrees of freedom:

$$84 + 112 - (33 + 17)^2/(21 + 20) = 135.02.$$

$S \times R : W$ adjusted for split plot regression with $12 - 1 = 11$ degrees of freedom:

$$112 - 17^2/20 = 97.55.$$

S adjusted for split plot regression is the difference of two adjusted sums of squares:

$$181.79 - 97.95 = 84.24.$$

$S \times W$ adjusted for split plot regression is a difference of two adjusted sums of squares:

$$135.02 - 97.55 = 37.47.$$

A table of means before adjustment for the covariate appears as Table 11.5. The various means adjusted for regression are given below.

Whole plot means of Y, $\beta_w = 4/1 = 4$:

$$\bar{y}_{1.}(\text{adj.}) = 6 - 4(2 - 2.5)/1 = 8$$
$$\bar{y}_{2.}(\text{adj.}) = 8 - 4(3 - 2.5)/1 = 6.$$

Split plot means of Y, $\beta_s = 17/20 = 0.85$:

$$\bar{y}_{..1}(\text{adj.}) = 6 - 17(2.5 - 2.5)/20 = 6.00$$
$$\bar{y}_{..2}(\text{adj.}) = 7 - 17(2 - 2.5)/20 = 7.425$$
$$\bar{y}_{..3}(\text{adj.}) = 4 - 17(2 - 2.5)/20 = 4.425$$
$$\bar{y}_{..4}(\text{adj.}) = 11 - 17(3.5 - 2.5)/20 = 10.15.$$

Table 11.5. Whole Plot, Split Plot, and Split Plot by Whole Plot Means.

	S1		S2		S3		S4		Mean	
	Y	Z	Y	Z	Y	Z	Y	Z	Y	Z
W1	5	2	8	3	4	1	7	2	6	2
W2	7	3	6	1	4	3	15	5	8	3
Mean	6	2.5	7	2	4	2	11	3.5	7	2.5

Interaction means $\bar{y}_{.ij}$ adjusted for both whole plot and split plot regressions:

$$\bar{y}_{.11}(\text{adj.}) = 5 - 4(2 - 2.5) - 0.85(2 - 2) = 7.00$$
$$\bar{y}_{.12}(\text{adj.}) = 8 - 4(2 - 2.5) - 0.85(3 - 2) = 9.15$$
$$\bar{y}_{.13}(\text{adj.}) = 4 - 4(2 - 2.5) - 0.85(1 - 2) = 6.85$$
$$\bar{y}_{.14}(\text{adj.}) = 7 - 4(2 - 2.5) - 0.85(2 - 2) = 9.00$$
$$\bar{y}_{.21}(\text{adj.}) = 7 - 4(3 - 2.5) - 0.85(3 - 3) = 5.00$$
$$\bar{y}_{.22}(\text{adj.}) = 6 - 4(3 - 2.5) - 0.85(1 - 3) = 5.70$$
$$\bar{y}_{.23}(\text{adj.}) = 4 - 4(3 - 2.5) - 0.85(3 - 3) = 2.00$$
$$\bar{y}_{.24}(\text{adj.}) = 15 - 4(3 - 2.5) - 0.85(5 - 3) = 11.30.$$

The variance of a difference between two adjusted whole plot means (11.15) is $V(\bar{y}_{.1.}(\text{adj.}) - \bar{y}_{.2.}(\text{adj.})) = 0[2/3(4) + (2-3)^2/1] = 0$, as E_a was zero for this example. The variance of a difference between two adjusted split plot means (11.16) is $V(\bar{y}_{..1}(\text{adj.}) - \bar{y}_{..2}(\text{adj.})) = 8.87[2/2(3) + (2.5 - 2)^2/20] = 3.07$. The variance of a difference between two adjusted split plot means at the same level of a whole plot (11.17) is $V(\bar{y}_{.11}(\text{adj.}) + \bar{y}_{.12}(\text{adj.})) = 8.87[2/3 + (2-3)^2/20] = 6.36$. The variance of a difference for two adjusted split plot means for the same level of the split plot treatment (11.18) is $V(\bar{y}_{.11}(\text{adj.}) - \bar{y}_{.21}(\text{adj.})) = 2(8.87 + 0)/2(3) + 0(2-3)^2/1 + 8.87(2 - 3 - 2 + 3)^2/20 = 2.96$.

A computer code in SAS GLM is given in Appendix 11.1. Not all the computations given above are obtainable by the code, for example some of the adjusted means.

Example 11.2: Covariate constant over split plots—The data for this example are those given by Miles-McDermott et al. (1988). The example was selected to be small so that the reader may perform the computations manually. A SAS code for these data is given in Appendix 11.2. The two whole plot treatments, W1 and W2, are arranged in a completely randomized design with four subjects per whole plot treatment. The two split plot treatments were randomized within each whole plot, resulting in a total of 16 observations. The data are presented in Table 11.6. The unadjusted whole plot means, split plot means, and split plot by whole plot totals appear in Table 11.7. An ANCOVA table is given in Table 11.8. Note that the sums of squares for Z and the cross products of Y and Z for factor S, split plot treatments, the $S \times W$ interaction, and Error S are all zero. This is because the covariate values are constant over all values of split plot treatments. This means that the split plot regression coefficient will be zero and will not enter into the adjustments of the split plot treatment means for this data set. Thus, only the whole plot regression coefficient is used in the adjustment of means. Likewise, the variances of differences between means will need to be changed to account for the zero split plot regression.

Error A sum of squares of Y, adjusted for the covariate with 5 degrees of freedom:

$$227.88 - (163.00)^2/159.50 = 227.88 - 166.58 = 61.30.$$

Table 11.6. Split Plot Experiment Design for Two Whole Plot Treatments, W1 and W2, in a Completely Randomized Design with r Replicates and with Two Split Plot Treatments, S1 and S2, Where the Covariate is Constant Over Split Plots.

Whole plot	Subject	Split plot S1 Y	Split plot S2 Y	Z	Total Y	Z
W1	1	10	8	3	18	6
	2	15	12	5	27	10
	3	20	14	8	34	16
	4	12	6	2	18	4
W2	5	15	10	1	25	2
	6	25	20	8	45	16
	7	20	15	10	35	20
	8	15	10	2	25	4
	Total	132	95	39	227	78
	Mean	16.5	11.9	4.88	14.19	4.88

W sum of squares of Y, adjusted for the covariate with 1 degree of freedom:

$$(227.88 + 68.06) - (12.38 + 163.00)^2/(2.25 + 159.50) - 61.30 = 44.48.$$

Whole plot regression coefficient:

$$163.00/159.50 = 1.022.$$

Note that the unadjusted means for split plot means and split plot by whole plot interaction means are the same as the adjusted means since the adjustments are zero. The various means adjusted for the covariate are:

$$\bar{y}_{.1.}(\text{adj.}) = 12.12 - 1.022(4.50 - 4.88) = 12.51.$$
$$\bar{y}_{.2.}(\text{adj.}) = 16.25 - 1.022(5.25 - 4.88) = 15.87.$$

Table 11.7 Totals and Means for the Data in Table 11.6.

Variable	Split Plot S1 Y	Z	S2 Y	Z	Total Y	Z	Mean Y	Z
W1	57	18	40	18	97	36	12.12	4.50
W2	75	21	55	21	130	42	16.25	5.25
Total	132	39	95	39	227	78	14.19	4.88
Mean	16.50	4.88	11.88	4.88	14.19	4.88		

Table 11.8. ANCOVA for the Data in Table 11.6.

Source of variation	Degrees of freedom (DF)	Sums of products			DF	Adjusted sums of squares	Mean square
		yy	yz	zz			
Total	16	3609	1282	542			
Mean	1	3220.6	1106.6	380.2			
W	1	68.06	12.38	2.25			
Error A	6	227.88	163.00	159.50	5	61.30	12.26
S	1	85.56	0	0			
S × W	1	0.56	0	0			
S × R : W	6	6.38	0	0	6	6.38	1.06
W (adj. for β_w)					1	$W'_{yy} = 44.48$	44.48
S (adj. for β_s)					1	$S'_{yy} = 85.56$	85.56
S × W (adj. for β_s)					1	$I'_{yy} = 0.56$	0.56

$$\bar{y}_{.1}(\text{adj.}) = 16.50 - 0(4.88 - 4.88) = 16.50.$$
$$\bar{y}_{.2}(\text{adj.}) = 11.88 - 0(4.88 - 4.88) = 11.88.$$
$$\bar{y}_{.11}(\text{adj.}) = 14.25 - 1.022(4.50 - 4.88) - 0(4.50 - 4.50) = 14.64.$$
$$\bar{y}_{.12}(\text{adj.}) = 10.00 - 1.022(4.50 - 4.88) - 0(4.50 - 4.50) = 10.38.$$
$$\bar{y}_{.21}(\text{adj.}) = 18.75 - 1.022(5.25 - 4.88) - 0(5.25 - 5.25) = 18.38.$$
$$\bar{y}_{.22}(\text{adj.}) = 13.75 - 1.022(5.25 - 4.88) - 0(5.25 - 5.25) = 13.38.$$

The estimated variances of a difference of two adjusted means are:

$$V(\bar{y}_{.1.}(\text{adj.}) - \bar{y}_{.2.}(\text{adj.})) = E_a[2/rb + (\bar{z}_{.1.} - \bar{z}_{.2.})^2/D_{zz}]$$
$$= 12.26[1/4 + (4.50 - 5.25)^2/159.50] = 3.06.$$
$$V(\bar{y}_{..1}(\text{adj.}) - \bar{y}_{..2}(\text{adj.})) = 2E_b/ra = 2(1.06)/4(2) = 0.26.$$
$$V(\bar{y}_{.11}(\text{adj.}) - \bar{y}_{.12}(\text{adj.})) = 2E_b/r = 2(1.06)/4 = 0.53.$$
$$V(\bar{y}_{.11}(\text{adj.}) - \bar{y}_{.21}(\text{adj.})) = 2[(b-1)E_b + E_a]/rb + E_a(\bar{z}_{.1.} - \bar{z}_{.2.})^2/D_{zz}$$
$$= 2(1.06 + 12.26)/4(2) + 12.26(4.50 - 5.25)^2/159.50 = 3.37$$

A SAS code and output of the program for aid in obtaining the above analyses are given in Appendix 11.2. Not all the desired computations are obtainable using this code. Some need to be computed manually.

11.3. COVARIANCE ANALYSIS FOR A SPLIT BLOCK EXPERIMENT DESIGN

The covariance analysis of variance described here is for a standard split block experiment design. There are three error terms in this design and hence three regression coefficients. Some means only require one regression for the covariate

Table 11.9. ANCOVA for a Standard Split Block Experiment Design.

Source	DF	Sum of products			DF	Adjusted sum of squares
Total	rab	T_{yy}	T_{yz}	T_{zz}		
Mean	1	M_{yy}	M_{yz}	M_{zz}		
Replicate, R	$r-1$	R_{yy}	R_{yz}	R_{zz}		
Factor A	$a-1$	W_{yy}	W_{yz}	W_{zz}		
Error A	$(a-1)(-1)$	A_{yy}	A_{yz}	A_{zz}	$ar-a-r$	$A_{yy} - \dfrac{A_{yz}^2}{A_{zz}}$
Factor B	$b-1$	U_{yy}	U_{yz}	U_{zz}		
Error B	$(b-1)(r-1)$	B_{yy}	B_{yz}	B_{zz}	$br-b-r$	$B_{yy} - \dfrac{B_{yz}^2}{B_{zz}}$
$A \times B$	$(a-1)(b-1)$	I_{yy}	I_{yz}	I_{zz}		
Error AB	$(a-1)(b-1)(r-1)$	C_{yy}	C_{yz}	C_{zz}	$(a-1)(b-1)(r-1)-1$	$C_{yy} - \dfrac{C_{yz}^2}{C_{zz}}$
Factor A adjusted for $\beta_a = A_{yz}/A_{zz}$		$a-1$				$W_{yy} - \dfrac{(W_{yz}+A_{yz})^2}{W_{zz}+A_{zz}} + \dfrac{A_{yz}^2}{A_{zz}}$
Factor B adjusted for $\beta_b = B_{yz}/B_{zz}$		$b-1$				$U_{yy} - \dfrac{(U_{yz}+B_{yz})^2}{U_{zz}+B_{zz}} + \dfrac{B_{yz}^2}{B_{zz}}$
$A \times B$ adjusted for $\beta_{ab} = C_{yz}/C_{zz}$		$(a-1)(b-1)$				$I_{yy} - \dfrac{(I_{yz}+C_{yz})^2}{I_{zz}+C_{zz}} + \dfrac{C_{yz}^2}{C_{zz}}$

adjustment, but others require all three coefficients. Five different variances of a difference between two adjusted means are required. The general situation is discussed first and this is followed by a numerical example illustrating the computational procedures. An ANCOVA table partitioning the degrees of freedom and sums of products appears in Table 11.9. The sums of squares for the variate Y adjusted for a covariate are also given in the table.

A liner model for data from a split block designed experiment with a covariate is as follows: (Federer, 1955, Federer and Meredith, 1992)

$$Y_{hij} = \mu + \rho_h + \alpha_i + \delta_{hi} + \beta_a(\bar{Z}_{hi.} - \bar{Z}_{...}) + \beta_j + \pi_{hj} + \beta_b(\bar{Z}_{h.j} - \bar{Z}_{...})$$
$$+ \alpha\beta_{ij} + \beta_{ab}(Z_{hij} - \bar{Z}_{hi.} - \bar{Z}_{h.j} + \bar{Z}_{...}) + \varepsilon_{hij}, \quad (11.23)$$

where $h = 1, \ldots, r, i = 1, \ldots, a, j = 1, \ldots, b$

Y_{hij} is the response for the hijth observation,

μ is a general mean,

ρ_h is the hth replicate random effect and is IID$(0, \sigma_\rho^2)$,

α_i is the ith factor A effect,

δ_{hi} is the random error effect of the hith observation and is IID$(0, \sigma_\delta^2)$,

β_j is the jth factor B effect,

π_{hj} is a random error effect of the hjth response and is IID$(0, \sigma_\pi^2)$,

$\alpha\beta_{ij}$ is the interaction effect of the ith level of factor A and the jth level of factor B, and ε_{hij} is the random error effect of the hijth response and is IID$(0, \sigma_\varepsilon^2)$.

The random effects, $\rho_h, \delta_{hi}, \pi_{hj}, \varepsilon_{hij}$, are assumed to be mutually independent.

β_a, β_b, and β_{ab} are the regression coefficients for factor A, factor B, and the interaction of A and B factors, respectively. Z_{hij} is the value of the covariate associated with the hijth experimental unit.

The formulae for computing the sums of cross-products have been presented above. Formulae for adjusting the means for the covariate are given in the discussion of the computational procedures for Example 11.3, as it is easier to see how the formulae are applied. Likewise, the general formulae for the variances of a difference between two means adjusted for a covariate are presented in the discussion of the numerical example. The expected value of Error A is taken to be $\sigma_\varepsilon^2 + b\sigma_\delta^2$, the expected value of Error B is $\sigma_\varepsilon^2 + a\sigma_\pi^2$, and the expected value of Error AB is σ_ε^2.

As is apparent from this example for a standard split block design, any variation that adds additional error terms will add additional regression coefficients and more complexity to the adjustment of means and the calculation of the variance of a difference between two adjusted means. These cases are not discussed herein but a straightforward extension of the methods discussed above should allow the analyst to obtain an ANCOVA for these situations.

Example 11.3: Covariate varies with smallest experimental unit—To illustrate the computational procedures for a standard split block experiment design with a covariate, a set of hypothetical data as given in Table 11.10 is used. The split block factors with factor A having two levels and factor B having three levels are each arranged in a randomized complete block experiment design. There are four complete blocks and thus $r = 4$ replicates.

Table 11.11 contains unadjusted totals and means needed to construct an analysis of covariance table. The ANCOVA table for the data of Table 11.10 is presented in Table 11.12. Following this table, it is shown how to adjust the means for factors A and B and the $A \times B$ interaction means. Then the variances of a difference between two means are presented. There are a total of five different variances that are required for comparing pairs of means. This demonstrates one of the complexities added when covariates are included in the statistical analyses.

Table 11.10. Responses for a Split Block Experiment Design with $a = 2$ Levels, A1 and A2, of Factor A, with $b = 3$ Levels, B1, B2, and B3, of Factor B, and with $r = 4$ Replicates for Levels of Factor A and for Factor B. Y is the Response Value and Z is the Value of the Covariate.

	Replicate 1				Replicate 2				Replicate 3				Replicate 4			
	A1		A2		A1		A2		A1		A2		A1		A2	
	Y	Z	Y	Z	Y	Z	Y	Z	Y	Z	Y	Z	Y	Z	Y	Z
B1	1	0	2	0	2	1	3	2	3	1	3	1	2	0	4	3
B2	2	1	2	2	3	2	3	3	3	1	2	1	2	0	3	2
B3	3	2	3	3	4	3	4	5	3	2	5	3	4	2	4	2

Table 11.11 Totals and Means for the Data of Table 11.10.

	A1		A2		Total			A1		A2		Total		Mean	
	Y	Z	Y	Z	Y	Z		Y	Z	Y	Z	Y	Z	Y	Z
R1	6	3	7	5	13	8	B1	8	2	12	6	20	8	2.50	1.00
R2	9	6	10	10	19	16	B2	10	4	10	8	20	12	2.50	1.50
R3	9	4	10	5	19	9	B3	14	9	16	13	30	22	3.75	2.75
R4	8	2	11	7	19	9	Total	32	13	38	27	70	42	2.92	1.75
Total	32	15	38	27	70	42									

	B1		B2		B3		Total		Mean	
	Y	Z	Y	Z	Y	Z	Y	Z	Y	Z
A1	8	2	10	4	14	9	32	15	2.67	1.25
A2	12	6	10	8	16	13	38	22	3.17	2.25

Means

	B1		B2		B3	
	Y	Z	Y	Z	Y	Z
A1	2.00	0.50	2.50	1.00	3.50	2.25
A2	3.00	1.50	2.50	2.00	4.00	3.25

The values for the estimated regression coefficients are $\beta_a = 0.67/1.67 = 0.401$, $\beta_b = 1.50/3.67 = 0.409$, $\beta_{ab} = 1.83/3.33 = 0.550$. The adjusted means for factor A treatments are:

$$\bar{y}_{.1.}(\text{adj.}) = \bar{y}_{.1.} - \beta_a(\bar{z}_{.1.} - \bar{z}_{...}) = 2.667 - 0.401(1.25 - 1.75) = 2.87.$$
$$\bar{y}_{.2.}(\text{adj.}) = \bar{y}_{.2.} - \beta_a(\bar{z}_{.2.} - \bar{z}_{...}) = 3.167 - 0.401(2.25 - 1.75) = 2.97.$$

The adjusted means for factor B treatments are:

$$\bar{y}_{..1}(\text{adj,}) = \bar{y}_{..1} - \beta_b(\bar{z}_{..1} - \bar{z}_{...}) = 2.50 - 0.409(1.00 - 1.75) = 2.81.$$
$$\bar{y}_{..2}(\text{adj.}) = \bar{y}_{..2} - \beta_b(\bar{z}_{..2} - \bar{z}_{...}) = 2.50 - 0.409(1.50 - 1.75) = 2.60.$$
$$\bar{y}_{..3}(\text{adj.}) = \bar{y}_{..3} - \beta_b(\bar{z}_{..3} - \bar{z}_{...}) = 3.75 - 0.409(2.75 - 1.75) = 3.34.$$

The adjusted means for the A by B treatment combinations are:

$$\bar{y}_{.11}(\text{adj.}) = \bar{y}_{.11} - \beta_a(\bar{z}_{.1.} - \bar{z}_{...}) - \beta_b(\bar{z}_{..1} - \bar{z}_{...}) - \beta_{ab}(\bar{z}_{.11} - \bar{z}_{.1.} - \bar{z}_{..1} + \bar{z}_{...})$$
$$= 2.00 - 0.401(1.25 - 1.75) - 0.409(1.00 - 1.75) - 0.550(0.50 - 1.25 - 1.00 + 1.75) = 2.51.$$

$$\bar{y}_{.12}(\text{adj.}) = \bar{y}_{.12} - \beta_a(\bar{z}_{.1.} - \bar{z}_{...}) - \beta_b(\bar{z}_{..2} - \bar{z}_{...}) - \beta_{ab}(\bar{z}_{.11} - \bar{z}_{.1.} - \bar{z}_{..2} + \bar{z}_{...})$$
$$= 2.50 - 0.401(1.25 - 1.75) - 0.409(1.50 - 1.75) - 0.550(1.00 - 1.25 - 1.50 + 1.75) = 2.80.$$

Table 11.12. ANCOVA for the Data of Table 11.10 with $r = 4$, $a = 2$, and $b = 3$.

Source of variation	DF	Sum of products			DF	Adjusted sum of squares
		YY	YZ	ZZ		
Total	24	224	142	108		
Mean	1	204.17	122.50	73.50		
Replicate, R	3	4.50	2.50	6.83		
Factor A	1	1.50	3.00	6.00		
Error A	3	0.50	0.67	1.67	2	$0.50 - 0.67^2/1.67 = 0.48$
Factor B	2	8.33	10.00	13.00		
Error B	6	1.00	1.50	3.67	5	$1.00 - 1.50^2/3.67 = 0.39$
A × B	2	1.00	0.00	0.00		
Error AB	6	3.00	1.83	3.33	5	$3.00 - 1.83^2/3.33 = 1.99$
Factor A adj. for $\beta_a = 0.67/1.67 = 0.401$					1	$1.50 - (3.00 + 0.67)^2/(6.00 + 1.67)$ $+ 0.67^2/1.67 = 0.013$
Factor B adj. for $\beta_b = 0.50/3.67 = 0.409$					2	$8.33 - (10.00 + 1.50)^2/(13.00 + 3.67)$ $+ 1.50^2/3.67 = 1.01$
Interaction adj. for $\beta_{ab} = 1.83/3.33 = 0.550$					6	$1.00 - (0.00 + 1.83)^2/(0.00 + 3.33)$ $+ 1.83^2/3.33 = 1.00$

$$\bar{y}_{.13}(\text{adj.}) = \bar{y}_{.13} - \beta_a(\bar{z}_{.1.} - \bar{z}_{...}) - \beta_b(\bar{z}_{..3} - \bar{z}_{...}) - \beta_{ab}(\bar{z}_{.11} - \bar{z}_{.1.} - \bar{z}_{..3} + \bar{z}_{...})$$
$$= 3.50 - 0.401(1.25 - 1.75) - 0.409(2.75 - 1.75) - 0.550(2.25$$
$$- 1.25 - 2.75 + 1.75) = 3.29.$$

$$\bar{y}_{.21}(\text{adj.}) = \bar{y}_{.21} - \beta_a(\bar{z}_{.2.} - \bar{z}_{...}) - \beta_b(\bar{z}_{..1} - \bar{z}_{...}) - \beta_{ab}(\bar{z}_{.11} - \bar{z}_{.2.} - \bar{z}_{..1} + \bar{z}_{...})$$
$$= 3.00 - 0.401(2.25 - 1.75) - 0.409(1.00 - 1.75) - 0.550(1.50 - 2.25$$
$$- 1.00 + 1.75) = 3.11.$$

$$\bar{y}_{.22}(\text{adj.}) = \bar{y}_{.22} - \beta_a(\bar{z}_{.2.} - \bar{z}_{...}) - \beta_b(\bar{z}_{..2} - \bar{z}_{...}) - \beta_{ab}(\bar{z}_{.22} - \bar{z}_{.2.} - \bar{z}_{..2} + \bar{z}_{...})$$
$$= 2.50 - 0.401(2.25 - 1.75) - 0.409(1.50 - 1.75) - 0.550(2.00$$
$$- 2.25 - 1.500 + 1.75) = 2.40.$$

$$\bar{y}_{.23}(\text{adj.}) = \bar{y}_{.23} - \beta_a(\bar{z}_{.2.} - \bar{z}_{...}) - \beta_b(\bar{z}_{..3} - \bar{z}_{...}) - \beta_{ab}(\bar{z}_{.23} - \bar{z}_{.2.} - \bar{z}_{..3} + \bar{z}_{...})$$
$$= 4.00 - 0.401(2.25 - 1.75) - 0.409(2.75 - 1.75) - 0.550(3.25$$
$$- 2.25 - 2.75 + 1.75) = 3.39.$$

Let the adjusted Error A mean squares be designated by $E_a = 0.48/2 = 0.24$, the adjusted Error B mean squares by $E_b = 0.39/5 = 0.08$, and the adjusted Error AB mean square by $E_{ab} = 3.00/5 = 0.60$.

The variance of a difference between two factor A means adjusted for a covariate is:

$$V(\bar{y}_{.1.}(\text{adj.}) - \bar{y}_{.2.}(\text{adj.})) = E_a \left[\frac{2}{br} + \frac{(\bar{z}_{.1.} - \bar{z}_{.2.})^2}{A_{zz}}\right] = 0.24[2/3(4)$$
$$+ (1.25 - 2.25)^2/1.67] = 0.184.$$

The variance of a difference between two adjusted means for factor B is:

$$V(\bar{y}_{.1}(\text{adj.}) - \bar{y}_{.2}(\text{adj.})) = E_b\left[\frac{2}{ar} + \frac{(\bar{z}_{..1} - \bar{z}_{..2})^2}{B_{zz}}\right]$$

$$= 0.08[2/2(4) + (1.00 - 1.50)^2/3.67] = 0.021.$$

The variances of a difference between two adjusted means for combinations of factors A and B are:

$$V(\bar{y}_{.11}(\text{adj.}) - \bar{y}_{.12}(\text{adj.})) = \frac{2[(a-1)E_{ab} + E_b]}{ar} + E_b\frac{(\bar{z}_{..1} - \bar{z}_{..2})^2}{B_{zz}}$$

$$+ E_{ab}\frac{(\bar{z}_{.11} - \bar{z}_{.12} - \bar{z}_{..1} + \bar{z}_{..2})^2}{C_{zz}} = 2[0.60 + 0.08]/8 + 0.08(1.00 - 1.50)^2/3.67$$

$$+ 0.60(0.50 - 1.00 - 1.00 + 1.50)^2/3.33 = 0.170 + 0.001 + 0 = 0.171.$$

$$V(\bar{y}_{.11}(\text{adj.}) - \bar{y}_{.21}(\text{adj.})) = \frac{2[(b-1)E_{ab} + E_a]}{br} + \frac{E_a(\bar{z}_{.1.} - \bar{z}_{.2.})^2}{A_{zz}}$$

$$+ \frac{E_{ab}(\bar{z}_{.11} - \bar{z}_{.21} - \bar{z}_{.1.} + \bar{z}_{.2.})^2}{C_{zz}} = 2[2(0.60) + 0.24]/12 + 0.24(1)^2/1.67$$

$$+ 0.60(0.50 - 1.50 - 1.25 + 2.25)^2/3.33 = 0.284.$$

$$V(\bar{y}_{.11}(\text{adj.}) - \bar{y}_{.22}(\text{adj.})) = 2\left[\frac{(a-1)E_{ab} + E_b}{ar} + \frac{(E_a - E_{ab})}{br}\right] + \frac{E_a(\bar{z}_{.1.} - \bar{z}_{.2.})^2}{A_{zz}}$$

$$+ \frac{E_b(\bar{z}_{..1} - \bar{z}_{..2})^2}{B_{zz}} + \frac{E_{ab}(\bar{z}_{.11} - \bar{z}_{.22} - \bar{z}_{.1.} + \bar{z}_{.2.} - \bar{z}_{..1} + \bar{z}_{..2})^2}{C_{zz}}$$

$$= 2[\{2(0.60) + 0.08\}]/8 + (0.24 - 0.60)/12] + 0.24(1.25 - 2.25)^2/1.67$$

$$+ 0.08(0.50 - 2.00 - 1.25 + 2.25 - 1.00 + 1.50)^2/3.33$$

$$= 0.170 - 0.030 + 0.144 + 0 = 0.284.$$

Note that the estimates of the variance components σ_δ^2 and σ_π^2 are negative and one would usually set them equal to zero, and the sums of squares are pooled to obtain single error variance. This was not done in the above variance computations as it was desired to show all the steps in computing the variances. The negative estimates account for the minus sign, -0.030, in the last variance above. Also, it is noted that some misprints occur in the variance formulae given by Federer and Meredith (1992). A SAS code for obtaining many of the above computations is given in Appendix 11.3.

11.4. COVARIANCE ANALYSIS FOR A SPLIT SPLIT PLOT EXPERIMENT DESIGN

In an ANCOVA for a standard split split plot designed experiment, there are three error variances and hence there will be three regression coefficients. A linear model for this type of design with a covariate is

$$Y_{hijk} = \mu + \rho_h + \alpha_i + \delta_{hi} + \beta_a(\bar{z}_{hi..} - \bar{z}_{....}) + \tau_j + \alpha\tau_{ij} + \varepsilon_{hij} + \beta_b(\bar{z}_{hij.} - \bar{z}_{hi..})$$
$$+ \gamma_k + \alpha\gamma_{ik} + \gamma\tau_{jk} + \alpha\gamma\tau_{ijk} + \pi_{hijk} + \beta_c(Z_{hijk} - \bar{z}_{hij.}), \quad (11.25)$$

where $h = 1,\ldots,r, i = 1,\ldots,a, j = 1,\ldots,b,$ and $k = 1,\ldots,c,$

Y_{hijk} is the $hijk$th response,

μ is a general mean effect,

ρ_h is the hth replicate effect and is IID$(0, \sigma_\rho^2)$,

α_i is the ith whole plot effect,

δ_{hi} is a random error effect associated with the hith whole plot and is IID$(0, \sigma_\delta^2)$,

β_a is the whole plot regression coefficient,

τ_j is the jth split plot effect,

$\alpha\tau_{ij}$ is the ijth whole plot-split plot interaction effect,

ε_{hij} is a random error effect associated with the hijth split plot and is IID$(0, \sigma_\varepsilon^2)$,

β_b is the split plot regression coefficient,

γ_k is the kth split split plot effect,

$\alpha\gamma_{ik}$ is the ikth whole plot-split split plot interaction effect,

$\gamma\tau_{jjk}$ is the jkth split plot-split split plot interaction effect,

$\alpha\gamma\tau_{ijk}$ is the ijkth three factor interaction effect,

π_{hijk} is a random error effect associated with the split split plot experimental unit and is IID$(0, \sigma_\pi^2)$, and

β_c is the split split plot regression coefficient.

The random effects, $\rho_h, \delta_{hi}, \pi_{hijk}, \varepsilon_{hij}$, are assumed to be mutually independent.

An ANCOVA for the above-mentioned response model is given in Table 11.13. The adjusted means are (Federer and Meredith, 1992) as follows:

$$\bar{y}_{.i..}(\text{adj.}) = \bar{y}_{.i..} - \beta_a(\bar{z}_{.i..} - \bar{z}_{....}) \quad (11.26)$$

$$\bar{y}_{..j.}(\text{adj.}) = \bar{y}_{..j.} - \beta_b(\bar{z}_{..j.} - \bar{z}_{....}) \quad (11.27)$$

$$\bar{y}_{...k}(\text{adj.}) = \bar{y}_{...k} - \beta_c(\bar{z}_{...k} - \bar{z}_{....}) \quad (11.28)$$

$$\bar{y}_{.ij.}(\text{adj.}) = \bar{y}_{.ij.} - \beta_a(\bar{z}_{.i..} - \bar{z}_{....}) - \beta_b(\bar{z}_{.ij.} - \bar{z}_{.i..}) \quad (11.29)$$

$$\bar{y}_{.i.k}(\text{adj.}) = \bar{y}_{.i.k} - \beta_a(\bar{z}_{.i..} - \bar{z}_{....}) - \beta_c(\bar{z}_{.i.k} - \bar{z}_{.i..}) \quad (11.30)$$

$$\bar{y}_{..jk}(\text{adj.}) = \bar{y}_{..jk} - \beta_b(\bar{z}_{..j.} - \bar{z}_{....}) - \beta_c(\bar{z}_{..jk} - \bar{z}_{..j.}) \quad (11.31)$$

$$\bar{y}_{.ijk}(\text{adj.}) = \bar{y}_{.ijk} - \beta_a(\bar{z}_{.i..} - \bar{z}_{....}) - \beta_b(\bar{z}_{.ij.} - \bar{z}_{.i..}) - \beta_c(\bar{z}_{.ijk} - \bar{z}_{.ij.}). \quad (11.32)$$

The estimated variances of a difference between two adjusted means for $i \neq i'$, $j \neq j'$, and $k \neq k'$ are given below:

$$E_a = \frac{A'_{yy}}{ar - a - r}, \quad E_b = \frac{B'_{yy}}{a(r-1)(b-1) - 1}, \text{ and } E_c = \frac{C'_{yy}}{as(p-1)(r-1) - 1}.$$

Table 11.13. ANCOVA for a Standard Split Split Plot Experiment Design Where the Covariate Z_{fhijk} Varies With the Split Split Plot Experimental Unit.

Source of variation	DF	Sums of products yy	yz	zz	DF	Adjusted sums of squares
Total	$rasp$	T_{yy}	T_{yz}	T_{zz}		
Mean	1	M_{yy}	M_{yz}	M_{zz}		
Block	$r-1$	R_{yy}	R_{yz}	R_{zz}		
Whole plot, W	$a-1$	W_{yy}	W_{yz}	W_{zz}		
Error A	$(r-1)(a-1)$	A_{yy}	A_{yz}	A_{zz}	$ra-a-r$	$A_{yy} - \dfrac{A_{yz}^2}{A_{zz}} = A'_{yy}$
Split plot, S	$s-1$	S_{yy}	S_{yz}	S_{zz}		
$W \times S$	$(a-1)(s-1)$	I_{yy}	I_{yz}	I_{zz}		
Error B	$a(r-1)(b-1)$	B_{yy}	B_{yz}	B_{zz}	$a(r-1)(b-1)-1$	$B_{yy} - \dfrac{B_{yz}^2}{B_{zz}} = B'_{yy}$
Split split plot, P	$p-1$	P_{yy}	P_{yz}	P_{zz}		
$W \times P$	$(a-1)(p-1)$	Q_{yy}	Q_{yz}	Q_{zz}		
$S \times P$	$(s-1)(p-1)$	U_{yy}	U_{yz}	U_{zz}		
$W \times S \times P$	$(a-1)(s-1)(p-1)$	V_{yy}	V_{yz}	V_{zz}		
Error C	$as(r-1)(p-1)$	C_{yy}	C_{yz}	C_{zz}	$as(r-1)(p-1)-1$	$C_{yy} - \dfrac{C_{yz}^2}{C_{zz}} = C'_{yy}$
W(adj. for β_a)	$a-1$					$W_{yy} - \dfrac{(W_{yz}+A_{yz})^2}{W_{zz}+A_{zz}} + \dfrac{A_{yz}^2}{A_{zz}} = W'_{yy}$
S(adj. for β_b)	$s-1$					$S_{yy} - \dfrac{(S_{yz}+B_{yz})^2}{S_{zz}+B_{zz}} + \dfrac{B_{yz}^2}{B_{zz}} = S'_{yy}$
$S \times W$(adj for β_b)	$(a-1)(s-1)$					$I_{yy} - \dfrac{(I_{yz}+B_{yz})^2}{I_{zz}+B_{zz}} + \dfrac{B_{yz}^2}{B_{zz}} = I'_{yz}$
P(adj. for β_c)	$c-1$					$P_{yy} - \dfrac{(P_{yz}+C_{yz})^2}{P_{zz}+C_{zz}} + \dfrac{C_{yz}^2}{C_{zz}} = P'_{yy}$
$W \times P$(adj. for β_c)	$(p-1)(a-1)$					$Q_{yy} - \dfrac{(Q_{yz}+C_{yz})^2}{Q_{zz}+C_{zz}} + \dfrac{C_{yz}^2}{C_{zz}} = Q'_{yy}$
$S \times P$(adj. for β_c)	$(s-1)(p-1)$					$U_{yy} - \dfrac{(U_{yz}+C_{yz})^2}{U_{zz}+C_{zz}} + \dfrac{C_{yz}^2}{C_{zz}} = U'_{yy}$
$W \times S \times P$ (adj. for β_c)	$(a-1)(s-1)(p-1)$					$V_{yy} - \dfrac{(V_{yz}+C_{yz})^2}{V_{zz}+C_{zz}} + \dfrac{C_{yz}^2}{C_{zz}} = V'_{yy}$

The various variances are (Federer and Meredith, 1992) as follows:

Estimated variance of a difference between two whole plot means adjusted for a covariate

$$V(\bar{y}_{.i..}(\text{adj.}) - \bar{y}_{.i'..}(\text{adj.})) = E_a\left[\frac{2}{rsp} + \frac{(\bar{z}_{.i..} - \bar{z}_{.i'..})^2}{A_{zz}}\right]. \quad (11.33)$$

Estimated variance of a difference between two split plot means adjusted for a covariate

$$V(\bar{y}_{..j.} - \bar{y}_{..j'.}) = E_b\left[\frac{2}{apr} + \frac{(\bar{z}_{..j.} - \bar{z}_{..j'.})^2}{B_{zz}}\right]. \quad (11.34)$$

Estimated variance of a difference for two split split plot means adjusted for a covariate

$$V(\bar{y}_{...k}(\text{adj.}) - \bar{y}_{...k'}(\text{adj.})) = E_c\left[\frac{2}{ars} + \frac{(\bar{z}_{...k} - \bar{z}_{...k'})^2}{C_{zz}}\right]. \quad (11.35)$$

Estimated variance of a difference between two means adjusted for a covariate for combinations ij and ij'

$$V(\bar{y}_{.ij.}(\text{adj.}) - \bar{y}_{.ij'.}(\text{adj.})) = E_b\left[\frac{2}{rp} + \frac{(\bar{z}_{.ij.} - \bar{z}_{.ij'.})^2}{B_{zz}}\right]. \quad (11.36)$$

Estimated variance of a difference between two means adjusted for a covariate for combinations ij and i'j

$$V(\bar{y}_{.ij.}(\text{adj.}) - \bar{y}_{.i'j.}(\text{adj.})) = \frac{2(\hat{\sigma}_\varepsilon^2 + \hat{\sigma}_\delta^2)}{rp} + \frac{E_a(\bar{z}_{.i..} - \bar{z}_{.i'..})^2}{A_{zz}}$$
$$+ \frac{E_b(\bar{z}_{.ij.} - \bar{z}_{.i..} - \bar{z}_{.i'j.} + \bar{z}_{.i'..})^2}{B_{zz}}. \quad (11.37)$$

Estimated variance of a difference between two means adjusted for a covariate for combinations ik and ik'

$$V(\bar{y}_{.i.k}(\text{adj.}) - \bar{y}_{.i.k'}(\text{adj.})) = E_c\left[\frac{2}{rs} + \frac{(\bar{z}_{.i.k} - \bar{z}_{.i.k'})^2}{C_{zz}}\right]. \quad (11.38)$$

Estimated variance of a difference between two means adjusted for a covariate for combinations ik and i'k

$$V(\bar{y}_{.i.k}(\text{adj.}) - \bar{y}_{.i'.k}(\text{adj.})) = \frac{2(s\hat{\sigma}_\delta^2 + \hat{\sigma}_\varepsilon^2 + \hat{\sigma}_\pi^2)}{rs} + \frac{E_a(\bar{z}_{.i..} - \bar{z}_{.i'..})^2}{A_{zz}}$$
$$+ \frac{E_c(\bar{z}_{.i.k} - \bar{z}_{.i..} - \bar{z}_{.i'.k} + \bar{z}_{.i'..})^2}{C_{zz}}. \quad (11.39)$$

Estimated variance of a difference between two means adjusted for a covariate for combinations ijk and ijk'

$$V(\bar{y}_{.ijk}(\text{adj.}) - \bar{y}_{.ijk'}(\text{adj.})) = E_c \left[\frac{2}{r} + \frac{(\bar{z}_{.ijk} - \bar{z}_{.ijk'})^2}{C_{zz}} \right]. \quad (11.40)$$

Estimated variance of a difference between two means adjusted for a covariate for combinations ijk and ij'k

$$V(\bar{y}_{.ijk}(\text{adj.}) - \bar{y}_{.ij'k}(\text{adj.})) = \frac{2(\hat{\sigma}_\varepsilon^2 + \hat{\sigma}_\pi^2)}{r} + \frac{E_b(\bar{z}_{.ij.} - \bar{z}_{.ij'.})^2}{B_{zz}}$$
$$+ \frac{E_c(\bar{z}_{.ijk} - \bar{z}_{.ij.} - \bar{z}_{.ij'k} + \bar{z}_{.ij'.})^2}{C_{zz}}. \quad (11.41)$$

Estimated variance of a difference between two means adjusted for a covariate for combinations ijk and i'jk

$$V(\bar{y}_{.ijk}(\text{adj.}) - \bar{y}_{.i'jk}(\text{adj.})) = \frac{2(\hat{\sigma}_\varepsilon^2 + \hat{\sigma}_\pi^2 + \hat{\sigma}_\delta^2)}{r} + \frac{E_a(\bar{z}_{.i..} - \bar{z}_{.i'..})^2}{A_{zz}}$$
$$+ \frac{E_b(\bar{z}_{.ij.} - \bar{z}_{.i..} - \bar{z}_{.i'j.} + \bar{z}_{.i'..})^2}{B_{zz}}$$
$$+ \frac{E_c(\bar{z}_{.ijk} - \bar{z}_{.ij.} - \bar{z}_{.i'jk} + \bar{z}_{.i'j.})^2}{C_{zz}}. \quad (11.42)$$

Under a fixed effects model, the expected value of E_a is $\sigma_\varepsilon^2 + p\sigma_\pi^2 + sp\sigma_\delta^2$, the expected value of E_b is $\sigma_\varepsilon^2 + p\sigma_\pi^2$, and the expected value of E_c is σ_ε^2. Also,

$$V(\bar{y}_{.ijk}(\text{adj.}) - \bar{y}_{.i'jk}(\text{adj.})) = V(\bar{y}_{.ijk}(\text{adj.}) - \bar{y}_{.i'j'k}(\text{adj.})) =$$
$$V(\bar{y}_{.ijk}(\text{adj.}) - \bar{y}_{.i'jk'}(\text{adj.})) = V(\bar{y}_{.ijk}(\text{adj.}) - \bar{y}_{.i'j'k'}(\text{adj.})),$$

and

$$V(\bar{y}_{.ijk}(\text{adj.}) - \bar{y}_{.ij'k}(\text{adj.})) = V(\bar{y}_{.ijk}(\text{adj.}) - \bar{y}_{.ij'k'}(\text{adj.})).$$

Some of the above variances use the estimated variance components. In these cases, the number of degrees of freedom will need to be approximated as explained previously. Most of the above variances without the covariate were given by Federer (1955) and in Chapter 3.

11.5. COVARIANCE ANALYSIS FOR VARIATIONS OF DESIGNS

As highlighted above, the analytical complexity of data from SPEDs and SBEDs increases when a covariate is added. Several formulae for obtaining the adjusted means were presented in the last section. Also, several different variances require computation when comparing the various means. The degrees of freedom for several of the variances are unknown and need to be approximated. For the experiment designs presented in Chapter 3 and the subsequent chapters, many formulae for computing the means adjusted for a covariate and for computing variances of a difference between two adjusted means will need to be obtained as they are not presently available. One approximate method to reduce the complexity is to use a series of simpler analyses.

11.6. DISCUSSION

In many situations, a covariance analysis is required in order to take care of the extraneous variation that can be controlled by covariates. For example, in running experiments on cultivars, the number of plants in an experimental unit may vary considerably owing to random differences in germination, weather conditions, and so forth. Random variation in the number of plants can be handled by covariance and should be used in comparing yield comparisons for various cultivars. If the experimenter expects to encounter this type of variation, it may be desirable to conduct a less complex experiment than some discussed in this book. More than one covariate may need to be considered in controlling extraneous variation. These are straightforward extensions of the results given in this chapter. The formulae for adjusting the means and computing the variances will need to be changed to accommodate situations as they arise in practice.

11.7. PROBLEMS

Problem 11.1. Factor A, $a = 5$ levels, is the whole plot factor, factor B, $b = 2$ levels, is the split plot factor and the design structure on the whole plot is a randomized complete block design with $r = 3$ blocks. The response variable of interest is Y, and Z is the covariate measured on the split plot experimental unit. The data are also included on the book's FTP site.

		Block					
		R1		R2		R3	
Factor A	Factor B	Y	Z	Y	Z	Y	Z
A1	B1	113	15	105	16	123	12
A1	B2	82	3	108	15	143	13
A2	B1	115	19	65	6	142	17
A2	B2	119	19	81	5	117	6
A3	B1	129	11	123	15	151	13
A3	B2	132	8	107	4	186	20
A4	B1	89	1	115	20	154	20
A4	B2	101	2	86	4	140	12
A5	B1	118	10	120	15	148	8
A5	B2	126	6	143	13	183	15

Conduct an analysis of covariance, obtain the adjusted means for Y, as well as the variances of a difference between two adjusted means using the data above.

Problem 11.2. Factor A with $a = 3$ levels, is the whole plot factor, factor B with $b = 5$ levels is the split block factor and the design structure on the whole plot is a randomized complete block design with $r = 4$ blocks. The response variable of interest is Y, and Z is the covariate Z measured on the smallest experimental unit. The data are included on the book's FTP site.

		Block							
		R1		R2		R3		R4	
A	B	Y	Z	Y	Z	Y	Z	Y	Z
A1	B1	246	19	220	13	186	4	246	19
A1	B2	207	14	172	5	174	6	157	2
A1	B3	133	11	118	7	99	2	122	8
A1	B4	173	16	178	17	128	4	140	8
A1	B5	153	6	158	7	177	12	165	9
A2	B1	218	7	247	14	201	3	248	14
A2	B2	225	14	198	7	179	2	226	14
A2	B3	138	7	149	10	147	9	123	3
A2	B4	171	11	162	8	196	16	193	16
A2	B5	230	20	203	13	192	11	179	7
A3	B1	200	10	180	5	196	9	187	7
A3	B2	157	4	170	7	197	14	177	9
A3	B3	86	2	133	13	115	9	142	15
A3	B4	133	8	132	8	174	18	171	18
A3	B5	163	11	127	2	172	13	184	16

Obtain an analysis of covariance, the adjusted treatment means, and the variances of a difference between two adjusted means using the data above.

Problem 11.3. The data for this problem are given on the book's FTP site. Write any additional code you need to obtain the means for A, B, and A*B treatments and variances of a difference between two means. A portion of the data and code are given below:

```
Data spex11_3;
input Y R A B Z W; /*Y=yield, R=block, A=planting method,
B=cultivation method, Z is the covariate, W is the covariate ijth
combination mean*/
X = Z - W;
datalines;
81.8 1 1 1 5 5.75
46.2 1 1 2 9 5.75
78.6 1 1 3 1 5.75
77.7 1 1 4 8 5.75
```

```
72.2 2 1 1 1 4.00
......
75.2 3 4 4 3 2.25
67.8 4 4 1 4 1.75
50.2 4 4 2 1 1.75
65.6 4 4 3 1 1.75
63.3 4 4 4 1 1.75
;

Proc glm data = spex11_3;
Class A B R;
model Y Z = R A A*R B A*B;
manova h = _all_/printh printe;
Lsmeans R A B A*B A*R;
quit;

Proc GLM data = spex11_3;
Class R A B;
Model Y = R A W A*R B A*B X/solution;
Run;
```

Problem 11.4. For the data of Example 11.3, insert a second covariate and obtain the analysis.

11.8 REFERENCES

Cochran, W. G. and G. M. Cox (1957). *Experimental Designs* (2nd edition). John Wiley & Sons, Inc., New York, pp 1–611.

Federer, W. T. (1955). *Experimental Design—Theory and Application*, Chapter XVI, Macmillan Publishing Co., New York, Chapter 16, pp 1–544.

Federer, W. T. (1975). The misunderstood split plot. In *Applied Statistics* (Proceedings of the Conference at Dalhousie University, Nova Scotia, May 2–4, 1974). R. P. Gupta, Editor. North Holland Publishing Co., Amsterdam, Oxford, pp. 9–39.

Federer, W. T. (1977). Sampling and blocking considerations for split plot and split block designs. Biometrische Zeitschrift (Biometrical Journal) 19:181–200.

Federer, W. T. (2003). Exploratory model selection for spatially designed experiments—Some examples. Journal of Data Science 1:231–248.

Federer, W. T., Z. D. Feng, M. P. Meredith, and N. J. Miles-McDermott (1987). Annotated computer output for split plot design: SAS GLM. Annotated Computer Output '87-8, Mathematical Sciences Institute, Cornell University, Ithaca, NY 14853.

Federer, W. T., Z. D. Feng, and N. J. Miles-McDermott (1987a). Annotated computer output for split plot design: GENSTAT. Annotated Computer Output '87-4, Mathematical Sciences Institute, Cornell University, Ithaca, N. Y. 14853.

Federer, W. T., Z. D. Feng, and N. J. McDermott (1987b). Annotated computer output for split plot design: BMDP 2V. Mathematical Sciences Institute, Cornell University, Ithaca, N. Y. 14853.

Federer, W. T. and H. V. Henderson (1978). Covariance analyses of designed experiments using statistical packages. Statistical Computing Section, Proceedings, American Statistical Association, pp. 332–337.

Federer, W. T. and H. V. Henderson (1979). Covariance analysis of designed experiments × statistical packages: An update. 12th Annual Symposium on the Interface, Computer Science and Statistics Proceedings, Waterloo, Ontario, May, pp. 228–235.

Federer, W. T. and M. P. Meredith (1989). Covariance analysis for split plot and split block designs and computer packages. ARO Report 89-1, Transactions of the Sixth Army Conference on Applied Mathematics and Computing, pp. 979–998.

Federer, W. T. and M. P. Meredith (1992). Covariance analysis for split-plot and split-block designs. *The American Statistician* 46(2):155–162.

Grimes, B. A. and W. T. Federer (1984). Comparison of means from populations with unequal variances. In *W. G. Cochran's Impact on Statistics* (editors: P. S. R. S. Rao and J. Sedrank), John Wiley and Sons, Inc., New York, pp. 353–374.

Kempthorne, O, (1952). *The Design of Experiments.* John Wiley & Sons, Inc., New York, Chapter 19.

Meredith, M. P., N. J. Miles-McDermott, and W. T. Federer (1988). The analysis of covariance for split-unit and repeated measures experiments using the GLM procedure (Part II). SAS Users Group International Conference Proceedings: SUGI 13, Cary, NC, SAS Institute Inc., pp. 1059–1064.

Miles-McDermott, N. J., W. T. Federer, and M. P. Meredith (1988). The analysis of covariance for split-plot and repeated measures experiments using the GLM procedure in SAS/STAT Software (Part I). SAS Users Group International Conference Proceedings: SUGI 13, Cary, NC, SAS Institute Inc., pp. 1053–1058.

Milliken, G. A. and D. E. Johnson (2002). *Analysis of Messy Data: Volume III. Analysis of Covariance.* Chapman & Hall/CRC, Boca Raton, London, New York, Washington D. C., Chapter 15.

Steele, R. G. D. and W. T. Federer (1955). Yield-stand analyses. Journal of the Indian Society of Agricultural Statistics 7:27–45.

Yates, F. (1937).The design and analysis of factorial experiments. Imperial Bureau of Soil Science Technical Communications, 35:1–95.

APPENDIX 11.1. SAS CODE FOR EXAMPLE 11.1.

```
data cov;  /*Data and code for Example 11.1*/
length R T S Y Z 3;
input R T S Y Z Ztotal;
cards;
1 1 1   3 1   6
1 1 2   4 2   6
1 1 3   7 1   6
1 1 4   6 2   6
1 2 1   3 2   8
1 2 2   2 0   8
1 2 3   1 2   8
```

```
1 2 4 14 4  8
2 1 1     6 2  8
2 1 2 10 2  8
2 1 3     1 0  8
2 1 4 11 4  8
2 2 1     8 4 12
2 2 2     8 1 12
2 2 3     2 3 12
2 2 4 18 4 12
3 1 1     6 3 10
3 1 2 10 5 10
3 1 3     4 2 10
3 1 4     4 0 10
3 2 1 10 3 16
3 2 2     8 2 16
3 2 3     9 4 16
3 2 4 13 7 16
;

proc glm data = cov; /*Sums of squares, crossproducts, and means*/
class R T S;
model Y Z = R T R*T S S*T;
manova h=_all_/printh printe;
lsmeans R T R*T S S*T;
quit;

proc glm data = cov;
class R T S;
model Y = R T R*T S S*T Z/solution;
lsmeans S ;   /*Other means adjusted manually*/
run;
```

APPENDIX 11.2. SAS CODE FOR EXAMPLE 11.2.

```
data spcov2; /*Example 11.2*/
input block whole split Y Z;
cards;
1 1 1    3 2
1 1 2    4 2
1 1 3    7 2
1 1 4    6 2
1 2 1    3 2
1 2 2    2 2
1 2 3    1 2
1 2 4 14 2
2 1 1    6 3
2 1 2 10 3
2 1 3    1 3
```

APPENDIX **265**

```
2 1 4 11 3
2 2 1  8 4
2 2 2  8 4
2 2 3  2 4
2 2 4 18 4
3 1 1  6 5
3 1 2 10 5
3 1 3  4 5
3 1 4  4 5
3 2 1 10 5
3 2 2  8 5
3 2 3  9 5
3 2 4 13 5
;
title 'Example 11.2';

proc glm data = spcov2;
class block whole split;
model Y Z = block whole block*whole split split*whole;
manova h = _all_/printh printe;
lsmeans block whole block*whole split split*whole;
quit;

proc glm data = spcov2;
class block whole split;
model Y = block whole block*whole split split*whole Z/solution;
lsmeans whole split split*whole;
run;
```

APPENDIX 11.3. SAS CODE FOR EXAMPLE 11.3.

```
data sbcov; /*sbcoc11.3*/
input block A B Y Z;
datalines;
1 1 1 1 0
1 1 2 2 1
1 1 3 3 2
1 2 1 2 0
1 2 2 2 2
1 2 3 3 3
2 1 1 2 1
2 1 2 3 2
2 1 3 4 3
2 2 1 3 2
2 2 2 3 3
2 2 3 4 5
3 1 1 3 1
3 1 2 3 1
```

```
3 1 3 3 2
3 2 1 3 1
3 2 2 2 1
3 2 3 5 3
4 1 1 2 0
4 1 2 2 0
4 1 3 4 2
4 2 1 4 3
4 2 2 3 2
4 2 3 4 2
;

proc glm data = sbcov;
class block A B;
model Y Z = block A block*A B block*B A*B;
manova h = _all_/printh printe;
lsmeans block A block*A B block*B A*B;
quit;

proc glm data = sbcov;
class block A B;
model Y = block A block*A B block*B A*B Z/solution;
lsmeans A B A*B;
run;
```

Index

adjusted Error *AB* mean square 254
adjusting means for the covariate 243, 253, 256
analysis of variance 3, 17, 40, 47, 64, 66, 68, 73, 74, 76, 77, 78, 79, 80, 81, 82, 83, 99, 102, 104, 107, 108, 110, 113, 122, 124, 126, 128, 129, 132, 133, 137, 155, 157, 173, 174, 177, 179, 189, 191, 193, 194, 203, 214, 215, 217, 218
analysis of variance world record 147
analysis of variance, ANOVA 10
ANCOVA 241, 244, 245, 248, 251, 252, 255, 256
ANOVA 11, 40, 49, 73, 99, 135, 136, 171, 189, 217
appropriate error term 26
appropriate model 88
augmented experiment design 113, 169, 188
augmented incomplete block design 180
augmented Latin square design 169
augmented Latin square or rectangle design 180
augmented lattice square design 169, 180
augmented randomized complete block design 170, 171, 180, 189, 191, 193, 194
augmented split block design 113, 169, 180, 188, 194
augmented split plot design 169, 180
augmented split split plot design 176, 180

augmented treatment 169, 189
axiom 114

balanced block 100
balanced block arrangement 76
Behrens-Fisher 215
biological clock 43, 81
blocking 2, 6, 197, 240
Bonferroni procedure 24

calendar clock 43, 78, 80, 81
carryover effect 29
check treatment 169, 171, 179, 194
checks 1
coefficient of variation 86, 88
competition 130
completely randomized design 4
complexly designed experiment 120, 135, 137, 147
component of variance 80
computer code 54, 170, 174, 180, 205
computer program 17, 240
conduct of an experiment 3, 137
confounding 6, 11, 49, 62, 76, 77, 88, 90, 98, 108, 125, 136, 153, 155
continuing effect 29
contrast 17, 88, 153, 155, 162, 173, 192
control treatment 111, 189, 191
criss-cross design 39

Variations on Split Plot and Split Block Experiment Designs, by Walter T. Federer and Freedom King
Copyright © 2007 John Wiley & Sons, Inc.

crossover design 29
cross-products 241, 245, 246, 252

damaged experimental unit 202
degrees of freedom 13, 45, 49, 65, 203, 243, 244, 259
design of the experiment 1, 137
direct effect 29
Dunnett's test 24, 25, 53

effective error variance 28
Error A 12, 44, 66, 91, 203
Error AB 44
Error B 12, 27, 44, 66, 67, 107, 214
Error C 67
error mean square 110, 111, 133, 218
error term 99, 108, 109, 138, 190, 193, 203
error terms 104, 122, 125, 127, 129, 135, 147, 155, 162, 191, 241
error variances 255
estimability 122
expected value 252, 259
expected value for factor B mean square 67
expected value of mean squares 13, 43, 44, 66
experiment design, definition 1
experimental unit 4
experimental unit, definition 2
experiment-wise error rate 24
exploratory model selection 3, 62
extraneous variation 260

factorial, definition 1
factorial design 39, 73, 90, 108, 109, 110
field design of the experiment 147 *et seq.*
fixed effect 9, 13, 14, 54, 64, 66, 67, 170, 173, 174, 216, 259
fractional factorial 91
fractional replicate, definition 1
F-statistics 17, 204, 216, 218, 241
F-test 3, 12, 44, 50, 51, 66, 67, 76, 105, 129, 132, 179, 203, 205, 218
F-values 47, 194

GENDEX 1
GENSTAT 239
geometrical components of an interaction 110
gradients 170

greater mean square rule 86
group of experiments 219
Guinness Worlds Records 147

half normal probability plot 133
honestly significant difference 24, 53

incomplete block design 4, 61, 74, 75, 98, 105, 107, 113, 189, 190
interaction 18, 22, 27, 44, 48, 89, 110, 192
interaction of treatment with covariate 240
intercropping 7, 188, 193, 197
inter-plot competition 170, 180
intrablock error 107, 190

key-out of degrees of freedom 109, 110, 123, 135, 137

Latin square design 4, 9, 47, 77, 97, 98, 100, 102, 104, 123, 124, 125, 153
least significant difference 52
linear model 9, 67, 99, 101, 121, 125, 128, 170, 190, 192, 214, 217, 251, 255, 256
linear model for split split plot experiment design 63
linear response model for the standard split plot experiment design 9
local control 2
log transformation 86

means adjusted for the covariate 249
measure of the efficiency 30
measure of the precision 54
missing observations 27, 202, 205
mixed effect 14, 54, 67, 251
mixed model 73
mixing effects 193, 197
mixture experiment 193
multiple comparisons 23, 26, 52, 138, 216
multiple error terms 239, 240
multiple range tests 3
multiple regression 240
multivariate analysis 239
new treatment 113, 169, 171, 179, 189, 191, 194

observational error 80
observational unit, definition 2

orthogonal 49, 54, 216, 218
orthogonal, pairwise 122
orthogonality 6, 108, 125, 129, 205
orthogonality, definition 122

parsimonious design 170, 180, 188
partitioning of degrees of freedom 9, 27, 41, 64, 68, 73, 74, 76, 77, 78, 79, 80, 81, 82, 83, 98, 102, 103, 107, 108, 113, 122, 126, 128, 132, 133, 137, 151, 155, 157, 170, 173, 177, 179, 189, 190, 192, 193, 203, 214, 215, 217, 251
per comparison error rate 24
permanent effect 29
placebos 1
points of reference 1
polynomial regression model 239
pooling procedures 133
precision estimates 53, 54
precision of the contrasts 29
PROC MIXED 54

random effect 13, 14, 27, 64, 66, 67, 68, 102, 121, 126, 128, 173, 174, 197, 216, 241, 256
randomization 2, 4, 39, 62, 63, 67, 69, 109, 111, 121, 122, 125, 127, 133, 135, 137, 180, 189, 193, 240, 244
randomization procedure and the layout of a split block design 40
randomized complete block design 3, 7, 8, 10, 39, 40, 41, 46, 47, 62, 77, 79, 80, 97, 98, 102, 104, 108, 113, 120, 125, 127, 135, 137, 190, 215, 218, 240, 244, 252
randomization plan for a standard or basic split plot 4, 5, 6
regression model 240
repeated measurements 29, 43
replication 2, 98, 137, 155, 157, 169, 170, 173, 174, 177, 189, 215, 252
residual effect 241
residuals 138
response model 3, 43, 137
response model equation with a covariate 241
row-column design 61, 75, 97
row-column error mean square 91

sample units, definition 2
sampling error 80
sampling units, definition 2
SAS 112, 113, 129, 171, 172, 180, 196, 218, 219
SAS code 47, 90, 105, 171, 248, 250
SAS PROC GLM 49, 50, 88, 91, 99, 104, 107, 129, 171, 173, 194, 195, 196, 202, 203, 213, 214, 215, 216, 218, 248
SAS PROC MIXED 73, 107, 138, 173, 174, 194, 213, 216
Scheffe's procedure 24, 25
screening 170, 188, 197
single degree of freedom contrasts 133
software packages 205, 239
spatially designed experiments 91
split block designs with controls 188
split block design 39
split block split block design 100
split plot design 3
split plot experimental unit 3
split plot regression 241, 247, 248
split split plot design 61, 62
split split split plot design 61, 67
square root transformation 86
standard error of a difference 14, 15, 16, 23, 45, 51, 64, 65, 69, 70, 71, 138, 171
standard or control 1, 3, 113
stratification 2
strip block design 39
strip strip block design 100
strip-plot design 39
sub-treatments in strips across blocks 39
systematic arrangement 61, 100, 132

three-factor factorial 11, 121, 125
three-factor interaction 90
time periods 62
transformation 215, 218
treatment design 3, 121
treatment design, definition 1
treatment predictor variable 239, 240
treatment variable 240
treatment, definition 1
trend analysis 240
triple lattice incomplete block design 6, 17
Tukey's one-degree-of-freedom sum of squares 11, 12

Tukey's studentized multiple range test 24, 53, 213, 216
two-factor factorial 3
two-way whole plot design 39

valid estimate of an error variance 3
variance component 3, 13, 73, 244, 255, 259
variance heterogeneity 215, 218
variance of a difference 46, 66, 71, 72, 250, 255

variance of a difference between two adjusted whole plot means 248
variance stabilizing transformation 215
variances of a difference between two adjusted means 243, 252, 258, 259

whole plot experimental unit 3, 39
whole plot regression 241, 246, 248, 249

Youden design 75, 76, 100

CPSIA information can be obtained
at www.ICGtesting.com
Printed in the USA
BVOW06*1656170417
480867BV00013B/86/P